PRAISE FOR
CREATIVE ALCHEMY: THE SCIENCE OF MIRACLES

'I am totally blown away by your book! It is a must for anyone to read, can't wait until it is available so I can recommend/buy it for all my clients, family and friends. Thank you so much for writing it. It is a masterpiece.'
– *Brigitte Djie, Master Coach for Tony Robbins*

'In her new book, *Creative Alchemy: The Science of Miracles,* Renaissance woman Stephanie Sinclaire Lightsmith offers you a powerful system for becoming the best version of yourself – the self many of us have long dreamed of being. In this visionary book, you will be guided on a path toward personal transformation many of us are seeking. It is an invitation to cultivate a new way of being that will free you from entrenched habits that cause fear, anxiety, and stress and help you experience balance and well-being in your mind, body and in relationships. Part I provides a solid foundation of understanding that draws from a wide range of sources including cutting-edge neuroscience research and the wisdom of ancient spiritual masters. Part II provides a "blueprint" of the Creative Alchemy system Lightsmith has taught in numerous workshops around the world over the past two decades. In those workshops, Lightsmith has helped adults and young people experience inner peace and joy. How fortunate we are that her able hand is now extended through this book and available to anyone who wants to grasp it.'
– *Dr Rebecca Weiss-Vlasic, clinical psychologist*

'I wholeheartedly recommend Stephanie Lightsmith's book *Creative Alchemy: The Science of Miracles.* This is an empowering book that can change your life! It contains pearls to inspire you to improve your life and the means to do it – to heal your relationship with your emotions and your past, manifest your dreams through directed willpower and imagination, and enhance your connection to your Inner Presence. Stephanie – an artist, entrepreneur, stage and film writer and director, explains how we can use specific tools to radically improve our lives and demonstrates how use of these tools is now supported by both leading-edge science and sacred knowledge, sharing a fascinating account of modern scientific studies and wisdom from ancient texts. The alchemy is transmuting our personality and story into gold for the good of all. I've experienced this alchemy, it really works!'
– *Gabrielle Meech, former human rights lawyer, teacher, RTT therapist*

'This book is a profound, accessible and compelling distillation of an amazing life experience combined with meticulous research and a genuine desire to help. ...ial that has the potential to change the trajectory of your life and the planet. Many books promise this, but none provide such a combination of watertight scientific evidence, passion and wisdom. Stephanie is an essential voice and I will be recommending this book to my therapy clients and my children. The way to a new planet is held within these pages. I can't recommend this enough.'

– *Malcolm Wilson, clinical hypnotherapist*

'I was truly impressed with the scope and magnitude of the new book, combining your life's journey with wisdom plus creating a step by step guide to realizing one's true inner self. Your magnificent technique is that your book is both a very captivating narrative of a most interesting and well-traveled life, at the same time a step-by-step guide to finding a person's own true center. You have captured and distilled the ancient wisdom of the ages and woven it with the most current scientific understanding in a most unusual and delightful manner. The fact that your method has been tested and proven in so many places worldwide, over time, by so many different people lends major efficacy to your important message to the people of today. Your personal knowledge is kaleidoscopic and vast, your life experiences and lessons learned reinforce the step-by-step process. The combined process brings the reader into the proper view through the guided example with strong reinforcing explanations. I always remember what professor Joseph Campbell used to say when asked, "What should someone do if they really want to help people and make a difference in this world?" His answer was very clear, "Teach people how to live in this world!" YOUR new book does this … so very well.'

– *Bob Davidow, entrepreneur and teacher/leader*

'Creative Alchemy's technique of Living Vision will expand your mind to the limitless realms of spontaneous creativity bringing the manifestation of your life into alignment with your highest purpose. The content is so rich and so enlightening and is going to benefit so many people.'

– *Hong Curley, Chinese medicine doctor, international business success coach, bestselling author, founder of the Quantum Healing Retreat and Aspire Humanitarian Foundation*

'A unique and powerful curriculum, designed by the Ascended Realms and our Daughter of Light, Stephanie, to raise consciousness – so all will embrace their missions with ease and grace. The Time is Now.'

– *Ascended Master Kuthumi and Dr Norma J Milanovich, author, worldwide speaker, and channel for the Ascended Realms*

PRAISE FOR THE TECHNIQUES OF
CREATIVE ALCHEMY: THE SCIENCE OF MIRACLES

'Stephanie intuitively knew exactly what I needed, she guided me through a cathartic and expressive creative process that reconnected me to a part of myself I never knew existed. Stepping through the layers she helped me embrace this powerful part of myself I'd abandoned. Since our session together I've felt calmer about being right where I am, while also noticing how things seem to be moving forwards more easily and joyfully for me. I can't recommend her highly enough, she's amazing!'

– *Penelope Beale, BodyIntuitive coach*

'I had the privilege to witness the extraordinary ways Stephanie expressed her divine talent, not just as a spiritual guide, but also as an intuitive teacher, an extraordinary artist, and a pure channel of love and devotion. Stephanie co- facilitated my Quantum Healing Retreats; her angelic presence graced every participant in ways that will never be forgotten. She helped each and every one of them with pure love and compassion; she helped the lost souls find their true north again; she helped to heal the traumas for those who lived with the memories of the past abuse and help them find love through forgiveness. Her Creative Alchemy workshops had created very dramatic impacts on our group of different ages. Everyone was transformed under her intelligent guidance. Some found their passion through the artistic expression, some exposed their hidden fears and limits, some broke free from their emotional prisons and some connected with the lost sense of love again. She has my highest regards, respect, love and eternal gratitude.'

– *Mike Curley, founder of Quantum Success Academy, Aspire Humanitarian Foundation, founding chief science officer, technical director Metagenics/Healthworld Australia*

'Creative Alchemy meets you at whatever level you are on and allows healing.'
– *Gabrielle Meech, former human rights lawyer, teacher, RTT therapist*

'Wow, what a weekend – words cannot even do it justice, it was one of the best experiences and I would do it again in a heartbeat. I am so grateful from the bottom of my heart to have been part of the experience with you all! Mum and I returned home feeling so much more balanced and connected. I also learned so much and feel like my awareness has massively expanded.'

– *Celeste Skatchill, designer and owner of Studio C*

'Stephanie is a metaphysical DJ. She has access to the music and can help us find the waves of our own flow. Creative Alchemy is the opposite of dogma, it tailors itself completely to the unique needs of each individual. I believe Creative Alchemy as a therapy has the potential to permanently relieve the disabling way we have held harmful beliefs about ourselves, and I believe Creative Alchemy as a therapy has the potential to permanently relieve the experiences. By creating art directly out of pain, the creative alchemist is able to bypass the limiting structures of language and culture that inhibit most traditional talk therapies (Clinical Analysis, CBT, etc.) from getting anywhere near the root problem. The power within Creative Alchemy is to skip past the conscious, rational mind of thoughts and self-imposed narratives to dive deeply and directly to the traumas one incurs during the earliest stages of life, when the feeling of what is happening greatly outweighs one's ability to describe it. These traumas become a subconscious warden to the remainder of our lives, imprisoning us within our self-limiting beliefs and creating negative behavioral patterns that seem to force our lives down an unchosen path. Like real magic, the work of Creative Alchemy unearths the root issues contributing to one's dis-ease in such a way that the practitioner is their own therapist and, rather than transference, has the end result of falling in love with themselves in the most positive way possible. In other words, Creative Alchemy not only gives you the power to heal yourself but to truly and deeply appreciate the person you have become to allow this healing to happen.'

– *Josh Davidow, filmmaker of* Journey with Robert Thurman in Bhutan

'Stephanie helped me edit the manual when I ran my first Quantum Healing Retreat in 2016. It was a great success and transformed the lives of 52 souls. Stephanie's powerful and beautiful presence contributed greatly to the success of this retreat. She brought with her the expertise of Creative Alchemy. The results she created for attendees were mind-blowing. One of our students created the blueprint of a quantum water generator through the visionary process Stephanie facilitated. I witnessed the techniques of Creative Alchemy transform an angry, depressed, anxious young woman into a free, happy, wise person in 40 minutes.'

– *Hong Curley, Chinese medicine doctor, international business success coach, bestselling author, founder of the Quantum Healing Retreat and Aspire Humanitarian Foundation*

'I had the privilege of meeting Stephanie at a Quantum Healing Retreat in Queensland, Australia. Instantly I knew her presence was magical and that I could learn something from her. Stephanie held one of the classes to demonstrate the effectiveness of emotional release through art with her techniques of Creative Alchemy. At first I was a bit skeptical. Throughout the last few days I had been battling consistent neck pains and stiffness, which was really getting to me! So I thought, well let's see if this works. We had to create an artwork based on what emotions we were currently feeling as a form of expression. My emotions were anger, frustration and rage due to the intense neck pain that I could not remove even after multiple treatments, oils, stretching, heat packs and more. So, my artwork was basically a nuclear BOMB going off releasing all these emotions. I didn't expect to get so involved in the artwork and I almost couldn't stop. My pain and stiffness dropped by 50 percent instantly. I was really impressed. I have also employed the Creative Alchemy techniques of creating inspirational images of what I would like to attract in my life to good effect.

'In a subsequent class that Stephanie facilitated for young entrepreneurs, we used the Creative Alchemy technique to create a symbol that encapsulated our unique expression for our life's purpose. While pondering what image best describes my purpose on this planet, a flash of a spiral DNA vortex appeared in my mind. This tornado-like image first started from the base as a DNA strand that twisted upwards while unfolding and unraveling into a tornado/vortex spinning energy field, growing larger as it spiraled upwards. This image represented to me the evolution of our consciousness and DNA expression through a vortex energy field (the quantum field) that which we are all connected to. That night I had trouble sleeping, I experienced restlessness and my brain was busy thinking away as usual. As I finally started to fall back asleep around 3.30 AM in a semi-conscious state I received an image/idea about water and DNA healing/repair. It involved a frequency machine that changes the molecular structure of water, which in effect allows for DNA healing and expression. I obviously can't go into detail on how the water machine works as I am currently researching and developing the product to this day. Water is a very strong passion of mine, and also anything to do with healing or anti- aging and repair. I am now driven and inspired to create a water machine that can now cover both of my passions, which has lead me on a fantastic ever-growing and learning adventure on the power of water and consciousness. I am so grateful for this experience! Everyone must try this technique that Stephanie has developed and taught over many years. Who knows where it can take you?'

– *Alex Bastianon, physical health professional, gym owner, inventor*

Alex went on to win the $50,000 Aspire Foundation Award to develop this invention.

CREATIVE ALCHEMY
THE SCIENCE OF MIRACLES

STEPHANIE SINCLAIRE LIGHTSMITH

Grateful acknowledgment is made for permission to reprint 'The Guesthouse' by Rumi. Translation copyright © by Coleman Barks. Permission granted by Coleman Barks.

Grateful acknowledgment is made for permission to quote from the Bridge of Freedom library. Permission granted by Werner Schroeder.

Grateful acknowledgement is made for permission to use the Future Memory technique taught by Dr Norma Milanovich. Permission is granted by Dr Milanovich.

Grateful acknowledgement is made for permission to use Body Communication co-created with Christina Hagman. Permission is granted by Christina Hagman.

Cover photo © Helen Williamson.
www.helenwilliamson.co.nz

Disclaimer: The information provided in this book is designed to provide helpful information on the subjects discussed. This book is not meant to be used, nor should it be used, to diagnose or treat any medical or psychological condition. For diagnosis or treatment of any medical or psychological problem, please consult your own physician. The author is not responsible for any specific mental, emotional or physical health conditions that may require medical supervision and are not liable for any damages or negative consequences from any treatment, action, application or preparation to any person reading or following the information in this book. References are provided for informational purposes only and do not constitute endorsement of any websites or other sources.

It should be noted that it is only through your consistent and sincere application of the theories taught within this book that you will begin to achieve your aims. Self-responsibility is crucial. Now that we've cleared that up, let's get on with the job of finding your purpose and fulfilment and co-creating a new world. All examples have had names and locations changed and are an amalgam of stories in order to protect the privacy of individuals.

CREATIVE ALCHEMY PUBLISHING

DEDICATION

This book is sincerely dedicated to *you*. May you become all you can be, experience the joy of fulfilling your innate potential and manifesting your dreams. May you gain the tools you need through *Creative Alchemy: The Science of Miracles* to heal your emotions, accelerate your Living Vision and activate the secret science of miracles. The mysteries of the *I AM Presence* are revealed for you in an accessible way so that modern people can utilize these truly powerful techniques that were once reserved for the few to create a purposeful, joyful and abundant destiny full of everyday miracles. May we also join one another in co-creating the world we long for, the paradise our Earth has the potential to be.

CONTENTS

Part Two • The Practical Application of Creative Alchemy

Part Three • The Broader Application

FOREWORD

It has been said that when the student is ready, the teacher will appear. The fact that you're holding this book in your hands is proof of that saying. *Creative Alchemy: The Science of Miracles* combines the latest findings of neuroscience and quantum physics with ancient wisdom and metaphysics that have previously been hidden from the public eye, considered too powerful for those who had not spent their lives refining their minds and emotions. Presented together in this special book they create a cohesive and accessible picture of exactly how it is that we can now learn to use our minds and emotions to alter our reality and thus create a new, abundant, joyful destiny while, at the same time, working together to co-create a new story for our planet. This book, along with all the carefully researched science, contains personal stories that help to illustrate how this system of Creative Alchemy has been successfully applied by people just like you over a 25-year period.

Stephanie Sinclaire Lightsmith has taken the path of the great sages of the past and transformed it for the feet of today's modern seekers. The benefit to you is that it ensures your goals are aligned with your highest good and the highest good of all. In this way, you can be sure that your journey through life will yield meaning and purpose, which are necessary for true happiness. Stephanie is a forerunner in the study and application of the technology of emotion, having begun this work a quarter century ago. We now know that it's not thoughts that fuel the manifestation of our dreams, it's emotion. And Stephanie has created techniques for healing our suppressed emotions that have been proven successful in schools, businesses, private retreats, clinics and with entrepreneurs. And when our emotions are healed, we become the powerhouses we were meant to be.

Stephanie also demonstrates how we can accelerate and elevate our imagination into what she calls a powerful Living Vision, by directing our emotion and then adding the 'secret sauce' of the animating electricity of our 'inner superhero' (the powerful Presence within). This creates a fail-safe template for becoming the masters of our life and living the magnificent life we were destined to live. You are about to learn how the application of these techniques has assisted entrepreneurs, teachers, students and people from all walks of life to realize their dreams and manifest their full potential, permanently changing their lives for the better as well as the lives around them.

Understanding how we can become the masters of our lives, successful, innovative, and solution oriented, is one of the most important pursuits of our time. When we realize we are no longer victims to circumstance and truly grasp the power we hold to shape our lives, many of our personal and planetary problems will be cured. When we understand that we can create the universe we need with the special techniques contained in this book, we need no longer be victims – or predators. When we truly grasp that we are each unique with our purpose and our plan, we will only compete with ourselves to become the best we can be, and we will realize that true collaboration is the way forward, and is the next frontier for humanity.

The revelation of Creative Alchemy is that we are all deeply creative, and this reality is proven scientifically and psychologically in this book. This creativity is intrinsic to our nature and this provides us with a new vision of who we are and what we are capable of. The truth is we are powerful masters able to create a beautiful and abundant life of meaning and purpose. You will see as you read the pages that follow, *Creative Alchemy: The Science of Miracles* is a crucial book for our times.

JACK CANFIELD
Co-author of the bestselling *Chicken Soup for the Soul*® series, co-author of *The Success Principles*™ and a featured teacher in *The Secret*.

INTRODUCTION

Welcome to *Creative Alchemy: The Science of Miracles*, the science, technique and art of consciously creating your most rewarding destiny and achieving your highest potential. It is a manual for masters in training, designed to assist in the shift of the ages currently occurring. This shift is humanity's transition from 'thinking' humans, *Homo sapiens*, to 'illumined' or 'enlightened' humans, *Homo illuminatus*, capable of far more than we previously thought possible. It is time we become masters of our destiny. Our primary tools will be the technology of emotion, the activation of Living Vision and connection to the animating Presence, the Source of life within us.

Creative Alchemy has been tried and tested with profound results in four countries with people from many different walks of life and age groups, beginning in 1995 when it was first manifested, streaming in on a river of profound wisdom I would then need to check against prevailing science and psychology, some of which has just caught up. The power that manifested it has led me by the hand, often mysteriously, to share the system with groups large and small and with individuals at retreats, in schools, businesses, with entrepreneurs and clinics, for 25 years at this writing. It is a completely comprehensive system that could be called a modernization of the Great Work by which the lead of our psyche may be transmuted to radiant gold. The time has come to share it with a broad audience. The raising up of humanity is the key solution to the problems facing our world.

You can feel confident your time will be invested well. With sincere commitment and application powerful shifts can be made, emotions can be healed, imagination can be strengthened and lifted and the true ancient secrets of manifestation can be successfully and consistently applied. In being the guardian and midwife of this system my life has changed entirely and for the better. I wish this for you.

There are three main tenets which anchor the system, the *Technology of Emotion*, the elevation of ordinary imagination to *Living Vision* and the science of miracles, understanding and applying the power of the *I AM Presence*.

The Technology of Emotion is the practical understanding of how to release suppressed and traumatic emotion, with accompanying techniques. Applying the techniques will allow you to access and redirect your energy, inner power, life force and vitality. The goal is not to be free of emotion but to have a free flow of *energy in motion*. You will learn *Radical Acceptance* of your emotional experiences and responsibility for them. You will learn how to remove condemnation, reaction and defense from your automatic responses and how to experience your emotions fully and deeply while understanding their message and origin without blame or projection. With this, the alchemy begins and a shift in perception and alignment follows. You begin to see your life through a clear lens and your profound role in creating it starts to become apparent. You also release a massive amount of energy previously used unconsciously to suppress your unhealed emotions that becomes superpower for your life and the fuel for the manifestation of your true purpose and dreams.

The activation of vision transforms ordinary imagination into Living Vision, an activated blueprint with the power to manifest into physical reality, if all crucial steps are taken. You will learn that you have a choice of consciously positively imprinting the force fields that surround you by becoming fully aware of your energy signature and its impact and how mastering that activity can begin to quickly change your life. Exactly how that process works will be explained in a way that allows you to comprehensively understand the details. The more we understand the process the greater the master we become and the easier it is to change.

You will learn why it is best to begin with manifesting qualities, *states of being*, in your life and how altering your state of being creates an inevitable transformation of the reality you currently inhabit. Your state of being is determined by the beliefs and perceptions that form your character, which in turn determine your destiny. Character *is* destiny. However, everything is malleable and shifts are possible on all levels of your being. Creative Alchemy will provide you with the right tools in a way that is easy to understand and apply. Becoming masters of our destiny is the next step in humanity's evolution. It is the key to solving our problems and the problems of our world. To be a master of destiny is to unlock the treasure chest of the universe and reclaim your birth right as a creator being with the power to manifest what you can conceive. The power you will learn of in this book; the power and mastery you will claim with diligent application is the true limitless power that always flows towards our highest good, and accelerates when we are aligned with our true purpose. We are each unique so competition will no longer be a factor in your life as you become the best you that only you can be.

For the purposes of Creative Alchemy we will use the term I AM Presence or Presence to describe the aspect of yourself where your superpowers are stored along with the plan and purpose for your life and your individual keynote. This multidimensional aspect of our selves is one with the Presence of all humanity and one with the universal Presence, the super-consciousness that permeates, enfolds and animates all. Connection to your Presence is like coming home. It is the source of your life, your consciousness, your unconditional love, abundance, true power and illumined wisdom. Called by the ancients the Beloved Mighty I AM Presence or the Magic Presence, this esoteric wisdom was once for the few, so great was the risk of abusing its power. Creative Alchemy has been designed to bring you into perfect alignment where your true joy awaits. This alignment ensures that the power that consistent connection to the Presence unleashes will not be used for harm nor can it accidentally harm you if all the steps provided are taken.

Unraveling the science of miracles will ensure you understand why and how this process is possible in a practical and accessible way. Modern science, quantum theory, neuroscience, cymatics, biology, physiology, psychology, esoteric wisdom and metaphysics will be referenced to explain core principles in a straightforward and accessible way. When we better understand the nature and fabric of our reality, we can see that what we call miracles are within the boundaries of natural law. A greater understanding of what the world around us is comprised of relaxes our mind and prepares it to receive concepts that are not yet part of the status quo of our current consensus reality.

The ultimate goal is to release into your world all the wisdom, power, love, personal genius, inspiration and abundance you innately hold for your journey on our shimmering emerald and blue planet. This is achieved by clearing anything that stands between you and the manifestation of your true authentic self. By the end of this book you will understand precisely how your emotions affect your biology, how your biology affects your perceptions, how your perceptions shape your beliefs and how your beliefs impact your reality. Emotion + Perception + Belief = Your Reality. Emotion is the fuel; belief and perception are the 'story.' Together they encode the energy field, or 'screen' around you and the movie of your life appears.

Your biology, emotions, mind and energy field continually emit patterns almost like a musical score, depending on our mindset. These patterns form the architectural blueprints around which your reality coalesces. Beliefs based on past trauma are colored with buried fears, grief and anger operating just below the surface, often hidden. This subterfuge of buried emotion, perceptual triggers and traumatic memory can affect what you manifest, causing the Frankenstein version of your dream to emerge. This is also known as *self-sabotage*. Suppressed emotion sends out distorted signals that create distorted patterns in the energy fields around you. The resulting manifestation will be just the same as if a wonky blueprint was used for building a house.

If you are unaware and unconscious of the blueprint you are imprinting on the energetic force fields around you, the cause and effect of your process will elude you, causing frustration and a continual sense of defeat. Once you master the system of Creative Alchemy, hiccups or boulders in your outer reality are seen as clues and can be shifted by rooting out the cause in your inner world.

> *If you want to find the secrets of the universe,*
> *think in terms of energy, frequency and vibration.*
> – Nikola Tesla

Creativity is the nature of reality. Energy fields vibrate and oscillate, their frequencies create geometric patterns and these patterns manifest. To truly understand the nature and power of our remarkable creativity and learn how to apply it is revolutionary and evolutionary. *We are all completely creative.* Creativity is built into our very nature and the structure of our brains. From creativity we emerge, through creativity we evolve. Even in perfect stillness we create. The most powerful thing we can create is our own lives. Great opportunity awaits us. We can co-create the world we want to share, play our work, follow our dreams, manifest wondrous things and live a celebratory life while we obtain our master's degree from Earth university.

Alchemy is an important term which refers to the transmutation of one substance into another. To many it suggests images of wizards of old attempting to turn lead into gold. It has also come to mean the Great Work, the most meaningful path for any of us: the transmutation of the lead of our psyches into the gold of our infinite potential.

As you become clear and free from inner distortion, you become more *in alignment*. To be in alignment means there is harmony and congruence between your mind, emotions, body, energy field and Presence. Your dreams begin to manifest in greater alignment with your conscious conception which is in alignment with your true purpose. As a fledgling master of destiny, you have a new formula:

> *Vision + Will + Emotion +*
> *the animating electricity of the Presence + Action =*
> *Consciously Created Reality*

As we walk the path of the Great Work, we are on a treasure hunt for our true essence. We never condemn ourselves. We employ our tool of Radical Acceptance. We cultivate a sense of wholeness and completeness. There is never anything wrong. We become *emotioneers,* emotional engineers navigating the inner map of our emotions to make adjustments to our external experience. As your awareness grows you will learn how to reverse engineer your manifested reality, tracing back to still-hidden emotion causing any wobbles. Ultimately, the techniques in Creative Alchemy form a system of conscious inner regulation by which you may adjust your external reality.

Our unconscious thoughts and emotions can cause unbridled chemical reactions in our biology, forcing us into an unrelenting state of overdrive. Many of us live this way, addicts of our inner pharmacy that wildly dispenses stressful 'drugs' such as adrenaline, designed to rocket us out of extreme peril. We react to life either in *fight or flight* mode or *paralysis*.

Even when we are not in strong emotional states our emotions, perceptions and beliefs are distorting our reality in ways in which we are not aware. By using the techniques of Creative Alchemy, you will learn how to place your inner superhero, your Presence, in the driver's seat. You will begin to see your life more as a curriculum, and concepts of failure and destructive habits of self-recrimination fade and then disappear. Gradually the process becomes seamless. It's a treasure hunt and your authentic self is the prize. When you understand the process of creating your life and co-creating your world, life becomes an adventure.

As your emotions are healed you will become increasingly still. A pool of internal silence and serenity begins to build a calm clear lake within you. The noisy children of your psyche have been attended to. Qualities of internal joy and contentment, a sense of safety and surety, balanced confidence, open-heartedness and open-mindedness begin to suffuse your life and world. Confidence builds as the practice brings you more success and you come closer to being the captain of your ship, consciously charting your course and skilfully navigating the occasional storm or buffeting winds.

When your cup is full your attention will naturally begin to turn outwards to others. Your desires begin to align with your own highest good and also the highest good of all, *which are never opposed*. You become a force for good in your life and in the world, which is very rewarding and will give meaning and purpose to your life while taking nothing from the fulfilment of your truest deepest dreams, the dream and plan your Presence holds for you. Creative Alchemy provides a step-by-step 'go at your own pace' system. Your sincere commitment and focused intention are all that is needed to begin.

Gradually we can truly understand how fluid and flexible this life is. We are slowed-down waves of energy, sound and light, communicating through frequency, tuned into everything around us, wearing a biochemical suit which is the instrument through which we share our songs with the world.

It is time for you to live your abundant joyful life and take your place in the great orchestra of this world. We are living on a planet that is currently undergoing a significant shift in consciousness, despite some appearances to the contrary. Physics, philosophy, psychology, biology, medicine, neuroscience, quantum mechanics and metaphysics are slowly merging into one sublime body of knowledge fueling the burgeoning human potential movement. The mechanistic, objective and compartmentalized classical scientific worldview is giving way to the integrated, subjective, conscious and quantum unified view of life – and this shift is affecting all disciplines and schools of thought, including those which relate to us as individuals. There is a mass awakening, causing a profound shift in consciousness and our understanding of human ability. There is a groundswell of evidence that we are not a mechanistic system of parts. In fact, we are fluid beings, made of condensed energy, electrical and magnetic and always in flux. Modern science confirms what the ancient mystics held true. Not only are our bodies, our DNA and our minds far more malleable than we have previously thought, we live in a sea of energy, information and consciousness that responds to our attention and intention and *we are that energy.* Never before have we had so many tools with which to consciously manifest a life of meaning and purpose. *The life we were born to live.*

As you follow the path of Creative Alchemy, you will begin to find your bliss as your atomic vibratory rate quickens. Your ability to access the information of the quantum field accelerates. Cognition gives way to illumination. You can see a bigger picture and more of the pattern is revealed. Solutions and innovation come spontaneously. As your consciousness and awareness expand you positively affect the consciousness of those around you. Science has now proven that the planet has a magnetic field which individual and group consciousness can affect either positively or negatively.[1] This relatively recent evidence that the human magnetic field impresses itself on the Earth's unified field of energy lends credence to what the mystics and sages of different traditions have always professed: *to become personally illumined lifts the whole of humanity.*

Part One will introduce you to your potential power and sevenfold nature, examining our four earthly vehicles, our *quantum generator,* the emotional body; our *quantum navigator,* the physical body; our *quantum conductor,* the subtle energy body and our *quantum manifestor,* our mental body and imagination. You will be introduced to the three aspects of our cosmic body, the inner superhero where our superpowers are stored – our *I AM Presence,* our *Causal Body* and our *Higher Mind* (or Self). You will learn about the *technology of emotion,* the *science of decree,* infinitely more powerful than affirmation, the difference between imagination and *Living Vision* and the role your identity plays in the formation of your life. We will examine the very fabric of reality and its mathematical and musical nature so that you will absolutely understand why and how you have the power to become a master of destiny and create your own reality. You will learn about the *science of miracles* brought together in a complete, concise and accessible way tailored for modern seekers of human potential and mastery for the first time, after 25 years of study and successful application with masters. This is the most crucial piece of the puzzle and key to our evolution from Homo sapiens to Homo illuminatus, powerful creator beings capable of shaping our lives and co-creating a harmonious paradise of our beloved planet. Our future is boundless and it starts with you, right now.

Part One is woven through with important information, examples and insights to help alter your perceptions and prepare you for the application of techniques designed to transform you from cocooned caterpillar to magnificent butterfly. With effort and sincere application, you will be on the road to becoming solution oriented, high functioning, inspired, innovative, joyful, fluid and revitalized – the master of your life, a master of destiny. Getting the balance right takes time but when you achieve it, the results are similar to a plug finding its socket. The light comes on and the way becomes bright.

Part Two is The Creative Alchemy Method™, a manual for masters in training – the Practical Application of Creative Alchemy in four phases:

1. *The Technology of Emotion.* How to heal suppressed emotion and release life force and vitality through the alchemy of creativity. Suppressing our emotions can rob us of 80 percent of our available energy resource. These powerful techniques have been taught and facilitated for 25 years in four countries with profound and consistent results.

2. *Living Vision.* How to create Living Vision from your imagination and inner vision to form energetic blueprints imbued with life force and power capable of manifesting in the world of form. By correct use of your imagination you become a magician, a Mage.

3. *Connection to the Magic Presence.* How to connect to the animating life force of your I AM Presence, your Source of power, wisdom, love and abundance; your Source of life, your 'inner individualized God self' or superhero.

4. *Four Templates for manifestation.* Template for Joy. Template for Health, Youth and Beauty. Template for True Prosperity & Abundance. Template for Finding Your True Partner.

It is crucial to master the first template before proceeding to the others for reasons you will learn. With persistence, all will be victorious. It is worth every moment of time invested. It is the means by which you can become all you can be. It is a life path, a great adventure – the journey of becoming your true authentic self and the master of your destiny.

Part Three applies Creative Alchemy to business, education, parenting, planetary wellness, art, the animal and plant kingdoms and other aspects of our shared world. Included in education are steps to change the world in one generation through applying Creative Alchemy principles to our education system.

To become Homo illuminatus, an illumined being, is to live and act with the awareness that you are one with the Presence, which permeates everything – what Max Planck, the Nobel Prize-winning father of quantum mechanics, called *conscious and intelligent Spirit, the matrix of all matter.*[2] You are interconnected to the whole but also unique with your own part to play. We live in a holographic, musical, mathematical, frequency-based universe. Every fractal aspect contains the entire blueprint of the whole, more and more of which becomes available to you when you expand your awareness. When your consciousness becomes united with your Presence, which is in turn united with the infinite consciousness of the universal Presence, the creation of the masterpiece version of your life and the positive evolution of our world becomes inevitable.

The practice of Creative Alchemy supports your transformation from the moment you intend to focus your attention on this path. Say YES and your 'future you' will enfold your 'present you' in the *eternal now.* Your future self is healed, whole and free – no longer stressing about the past or filled with angst about the future. You are integrated, all aspects of your beautiful self, shadow and light, past and future held in unconditional love and Radical Acceptance. You've become your own ideal parent. You've found your greatest wound and beheld the mystery of your greatest strength. The journey gets easier. Life becomes a treasure hunt. Life becomes fun!

If this book has landed on your mental doorstep, you are ready to become a creative alchemist, the conscious creator of your life – to experience contentment, meaningful purpose, abundance and the true success we all wish for. You are ready to sing your true note loud and clear in the symphony of life and take your rightful place. You are ready to become a force for good in the world with all the great rewards that this will bring. And finally, you are ready to access your unique genius, your highest visionary capacity and your inner superpowers, *expressing your much-needed unique self.*

If you can commit to applying the formula offered in this book and accept the invitation to consciously create your reality, your future is boundless.

To learn more about how the system of Creative Alchemy came to be and my personal story, please jump to Chapter Two, Part Three, and then come back for Part One. Otherwise, let's begin!

PART ONE

THE ALCHEMY OF POWER

Power is how we release our gifts into the world, how we manifest our dreams and destiny. Power must be balanced by love and tempered with wisdom. Of love, wisdom and power, love is paramount. Power without love creates fear. Wisdom without love is cold. But love without power and wisdom can become obsession, depression and sadness.[3]

We see extremes of power in our world today. There are those who abuse it and those who shy away from it. Learning to wield true power, not power over another but the directed power of your life force and the superpower of your Presence, is a crucial element of Creative Alchemy.

1

OUR QUANTUM GENERATOR – THE POWER OF EMOTION

You have two ways to live your life, from memory or inspiration.
– Ihaleakala Hew Len PhD [4]

Our emotions are the *quantum generator* of our life. Our emotions radiate into and permeate the energy fields surrounding us, creating patterns that influence manifestation. The emotional body holds enormous power and can draw on our energy reserves until we are drained or, when healed and directed, utilize our energy effectively to fuel the creation of the dream of our life.

The *technology of emotion* is one of the defining pursuits of the 21st century. Understanding our emotions is a relatively new science but the results are startling. It wasn't until the mid-1990s that magnetic resonance technology demonstrated the impact our emotions had on our physiology, which completely changed our understanding of the role of emotion in our mental health and physical well-being.

Emotions emerge from the landscape of the body and are interpreted by the mind as sensorial expressions we call feelings. [5] Neuroscientist Antonio R Damasio has made the distinction that *emotions* are chemical reactions cascading through the body whereas *feelings* are our perceptual response to these chemicals. [6] The emotions themselves are electrical and chemical signals that occur in the body in response to either external or internal stimuli. [7] Perhaps a little less poetic than we would like to think of these precious forces that fuel the inspirations of great love, poetry, literature, theater, film, architecture, art, invention, science, symphonies and other wonders of our world. But this knowledge is very exciting from the point of view of understanding emotion and exactly how it impacts on us.

When our emotions are fluid they are *energy in motion*. We can experience them deeply as they flow through us and transform. The ability to consistently access clear positive emotion provides the fuel for the manifestation of our life and purpose. When our emotions are trapped, buried and suppressed, they build up pressure that can cause overreactions, explosions with heartbreaking consequences. For example, explosions of uncontrolled anger have become the norm in our unhealed world leading to destruction in relationships, communities and in the extreme, abuse, rape and war. Suppressed anger, anger turned inwards, causes depression. Our planetary suicide rates are higher than they've ever been. Unexpressed grief can cause us to 'expect the worst' or it can form the root of irrational fears. Hidden memory and buried trauma can form a gray lens over our perceptions, completely distorting how we see the world, affecting our motivation and self-worth. Memories can play out continually in our minds like mental torture bathing our bodies with caustic stress hormones and chemicals until the accompanying trapped emotions are released. Our neural circuitry can solidify, creating habitual emotional response, *biologically* reinforcing old perceptions and beliefs which have limited us.

The good news is we can create new patterns that establish new neural circuitry that will stimulate new patterns of behavior. Our emotions have a significant effect on how we perceive our reality, how we respond to it and how we create it. [8] Once we release the stuck and hidden emotions that derail our lives by causing destructive behavior patterns, we begin to experience greater harmony – a big key for successful manifestation – and a vast amount of previously suppressed energy becomes available. When trapped emotion is released we also release personal power, energy, vitality and pure life force. We come alive in ways we haven't previously experienced since we were children and some of us, never before.

Emotions exert a powerful influence over cognitive processes including attention, memory, perception, decision-making, organizing and coordinating brain activity. [9] We now know they affect every aspect of our being and can run the show for better or worse. We are blessed to live in an era where neuroscience has been able to confirm the malleable, responsive and subjective nature of our brains, while 'new' science and quantum theory have demonstrated the malleable, responsive and subjective nature of what we call reality. Nothing is fixed and with the right tools positive adjustments can be made. These are the tools Creative Alchemy offers you.

The mind–body matrix stores the entire history of our emotional life. Repetitive thoughts, behaviors and emotions create neural pathways like record grooves causing us to repeat reactions and behavior when we feel triggered hijacking personal power and sovereignty. [10] This is why habits are challenging to change. *Our actual physiology is altered by our emotions.* Thankfully, our brains can be remoulded and that process is called *neuroplasticity.* [11] When we release a suppressed emotion we form new neural pathways that shift our responses to the physical world. We may retain the memory but its emotional charge and control over us ceases.

The study of neuroscience, particularly its findings on neuroplasticity – how flexible our brains are – has confirmed that we are living art, masterpieces in flux – mouldable and re-mouldable, based on our thoughts and feelings, our focus and attention. The field of epigenetics has revealed that our external environment and our internal environment of thought and emotion can alter our DNA, challenging the concept of genetic determinism that so affected scientific and medical thought with rigid absolutism. [12] Recent scientific studies have shown that our DNA emits signals that affect the fields of energy around us. We are living in an interactive field, and our biology, which we now understand to include our thoughts and feelings, our energetic field and our chemistry can interact with and change the field that surrounds and permeates us and vice versa. [13]

Our Inner Pharmacopoeia

Neurotransmitters are chemical messengers, molecules that carry information between neurons, our brain and nerve cells, and the other cells of our body. They are part of a complex network within us that affects us on emotional, physical and mental levels. Our cells have receptor molecules, which engage with our neurotransmitters in an ongoing call and response. Stress chemicals such as adrenaline and cortisol are taken up by these receptor molecules on our cells when we experience or recall stressful events, or if stressful emotions are buried. These particular chemicals are part of our body system designed to alert us to extreme danger. Most of us are soaking in them all the time, causing us to be over-alert and fatigued at the same time. If it is continual an addictive pattern emerges. Negative thinking can actually make you an addict to the chemicals of your inner pharmacy!

You may find yourself stewing over past events or worrying about a future you feel you have no control over. This will cause the continual release of the chemical cascades related to the underlying triggering emotion. If you have suppressed anger, for example, you are likely to perceive the world through a veil of irritation, annoyance and impatience. Even if it never builds to an external explosion, you will be subject to inner states of anxiety. Your receptor molecules will be used to receiving the chemicals that are released when you are irritated, annoyed or angry. It is likely that you will find yourself in many situations that feed these receptors their favorite drug, epinephrine (adrenaline), in response to your inner addiction. You may even pick fights with people because an unconscious part of you feels at home being angry and your receptor molecules are nagging for their 'hit.' Your discordant energy field will rub up against and prickle others with similar fields as the physical law of attraction – like attracts like – comes into play.

With suppressed grief, the world becomes a melancholy place. That which you focus on in your world will tend to be the sad stories of humanity. You may fall into victimhood, an inner scenario of self-pity and even self-loathing that saps your motivation and vitality, bathing your body in chemicals that slow you down. Victims attract bullies that cement their feelings of low self-worth and also those stuck in 'helper mode' who feel the need to earn their place in the world and often enable those they are trying to help.

Candace Pert PhD was a neuroscientist and pharmacologist who, by her discovery of the opiate receptor and binding sites for endorphins in the brain, greatly expanded our understanding of the physical nature of emotions and their role in health and consciousness. She called her discoveries *new-paradigm physiology*, because the facts of science aligned with the latest thinking of the most cutting-edge healers in the world. [14] Pert significantly advanced our understanding of the human body–mind interface and our potential for activating our imagination, beliefs and expectations by releasing our suppressed emotions and stored memories. She dared to marry esoteric thought, the study of the imagination, quantum mechanics, energy fields, frequency and vibration and other ways of exploring consciousness, and related them to the body and mind in a much deeper context than had previously been accepted in mainstream neuroscience.

Pert was one of the earliest neuroscientists to maintain that the body *is* the subconscious mind. She discovered that all our cells are intelligent entities. Emotions are the actual physical responses of the receptor molecules on our cells. They are stimulated by the neuropeptides that bind to them. Each is associated with a different state of being. Various neuropeptides stimulate various feelings. For example, the neuropeptide dopamine creates the mental and emotional experience of bliss. [15] Physical responses to emotion are produced directly at the cellular level. When we discover we have become negative neuropeptide addicts in a circular call-and-response process, we need to interrupt the pattern if we want to change. Creative Alchemy was developed to fill that need. It takes a period of time to be able to make this kind of shift, [16] but relief can begin immediately.

There are many ways to retrain the receptor molecules to crave neuropeptides that create positive states, replacing prior addictive states that flood the body with alarm best saved for running from wild tigers, but the most effective way is to release the trapped emotion and buried trauma that continually trips the switch on those chemicals. When your emotions are healed your thought patterns automatically change and your actual biology changes. New positive grooves of behavior are formed. With that, your beliefs and perceptions about yourself and the world also shift. When your previously suppressed emotions are released your beliefs and perceptions will be based on the present, not on what you've experienced in the past. They will be innocent and fresh, uncolored by past trauma. You will see the world differently and it will respond to your new vision of it. Vision is the ability to perceive clearly in the present moment, including perceiving what seems at first hidden. This could include patterns in your life and patterns in your culture, with startling new wisdom and greatly enhanced intuition.

Once you are free of your old stories and habitual emotions you will automatically create new neural pathways. You can become addicted positively to chemicals also supplied by your inner biology and feel the frequent bliss bath of dopamine and other 'feel-good' chemicals. [17] The frequency and vibration emitted by your cells will also shift, including how they interface with other energy fields, exchanging different information which begins to magnetize positive synchronicities. Your cells are living beings under your guidance. They await your acknowledgment and loving command. You are the general of a great army which, when aligned, becomes a powerful force in your life and in the world.

When your suppressed emotions have been cleared, another bonus is that your inspiration and motivation increase dramatically. Your mind becomes an open landscape in which your true purpose can reveal itself. Your purpose is that which gives meaning to your life. When your inner slate is wiped clean, your entire energy field shifts. New opportunities are magnetized to you. New community will emerge. Family and friends may react strongly to the new you. We are made of energy. The greater the energy, the higher the frequency, the shorter and more energetic the wavelengths the faster the vibratory rate will be. This heightened energy field can uplift others or cause catharsis. Higher, faster vibratory rates can push up heavy suppressed thoughts and emotions. This can be uncomfortable and challenging, causing discomfort in those who are suppressing emotion. As you become clearer your level of compassion for yourself and others naturally increases. You will have gained insights and coping tools. When we are not being driven by upset emotions it is easier to have empathy for others.

When our energy field is strong and clear we entrain the field around us instead of the other way around. When your energy fields become synchronized, synchronicity will begin to operate in your life. This is when unexplained positive or lucky experiences begin to occur. They seem at first like small miracles but they are actually your environment responding to your new-found alignment and congruence on the unseen level of energy.

The release of buried emotion requires your courage, for it takes courage to revisit and let go of old patterns and to re-experience feelings that weren't completely expressed at the time of your initial trauma. Often our brains have worked hard to suppress the emotions of intensely traumatic memories.

Memory deficits are one of the symptoms of post-traumatic stress disorder (PTSD). [18] I believe our entire world is currently experiencing post-traumatic stress disorder to some degree.

Trauma can be the full-blown severe trauma of an abused child. It can be the mild but affecting trauma of ongoing stress created by growing up in a culture that has no safe means of expressing strong emotion. Growing up with parents who are also wounded and therefore unable to love unconditionally can cause you to hide parts of yourself considered unwanted, disruptive or shameful. Had your parents been free and healed, they could have held you strongly and lovingly as you bloomed in your true essence, purpose and authenticity, teaching you safe ways for expressing all of you. You can now do that for yourself.

Creative Alchemy offers these tools so future generations can have a different experience. You can be the one who breaks the chain of pain for your lineage and those you love. As a species, we can become more loving and tolerant. Very few of us have grown up experiencing unconditional love. The good news is that as you heal, expand and step into true power, you give permission for all those in your life to do the same.

When we live with suppressed emotion we become either bottled up and 'flat' or the opposite, easily triggered and overly heightened – or you might vacillate between the two. When these emotions are released, anxiety diminishes or disappears along with self-pity and victim consciousness. We can learn to communicate our needs without blame or projection. We are free to operate in the world with fluidity and able to determine where we are on our journey by objectively assessing our current reality against our inner emotional state. To correctly process an emotion at the time of trauma, you need to fully express your feelings in a safe and authentic way, owning your emotions. Many cultures have no protocol for this and, in fact, advocate suppression. Has anyone ever told you *don't rock the boat*? Most of us have heard this and live by a version of that maxim.

A great deal of personal power and energy is applied to the task of suppression of our emotions. Suppression of emotion is also implicated in psychological conditions such as borderline personality disorder (BPD) and obsessive–compulsive disorder (OCD). [19] Creative Alchemy operates from the premise that the suppression of emotion is the underlying cause of most psychological disorders and also many physical illnesses. At the very least, clearing suppressed emotion needs to be addressed when we become physically ill to ensure complete healing.

When you release a suppressed emotion, you unleash an enormous amount of energy, the amount that it took to lock that emotion in place. You get this life force back. You gain the fuel and power to affect your life. You become able to be present, available and in alignment. Alignment is when your mind, emotions, body and energy field are congruent. To be congruent means your thoughts, feelings and actions are operating harmoniously and not at odds with one other. When you are in harmony your Presence automatically begins to act in your life.

When you come into balance, a state of being that is characterized by clear and flowing emotions with discordant memories no longer tugging at your psyche and coloring how you perceive the world and clear positive thoughts, then accurate discernment is restored. Your inner guidance, intuition and gut instinct switch on and provide accurate feedback. The distinction between clouded discernment and accurate discernment is the difference between driving a car with a dirty windshield and driving a car with a clean windshield. Crashes can be avoided!

How is it we can have yet another partner with all the same issues as our last one, when the relationship seemed so perfect at first? How is it we have reorganized our entire life but in a short period of time all our old problems reappear? Poor discernment. Your discernment is overridden to the degree your emotions are suppressed. Until you clear out your psychic baggage, the law of attraction will draw to you what is unresolved within you. You will actually unconsciously choose what is wrong for you again and again. Frequency attracting like frequency is not philosophy, it's physics. When our heart releases suppressed pain, there is room for joy to emerge. Phew! It feels good. We start to see trouble coming a mile away and can consciously choose a new direction.

With your energy no longer engaged in suppression, inner buoyancy, renewed passion for life and greater inspiration begin to surface. Laziness and boredom disappear. Both these states are signifiers of repressed energy. Commitment and motivation come easily. Intention is the defining arrow pointing to the shore of your new life and the new you.

All that is suppressed by us controls us. We cannot suppress selectively. When we bury old anger, grief, longing, resentment, bitterness and fear, we also bury our capacity for joy and our personal vitality – the ability to feel and be truly alive. Our suppressed emotions are volatile and easily triggered. Like a band-aid on an active volcano, a mere word or glance can unleash a destructive river of molten lava on others and ourselves.

When buried emotion is expressed and released, you will feel a strong sensation in your body sometimes accompanied by a cathartic release. *Catharsis* means purifying or cleansing, which in this context could involve a display of raw emotion, such as crying or laughing, or experiencing a pure, strong, physical sensation. These powerful sensations are what we have been afraid of experiencing. I call catharsis *pain on the way out* because enormous relief follows it. All our emotions begin to flow like a river. This flow is deeply life-enhancing and affects all systems of the body, subtle energy fields and mind. The lead of heavy feeling is transformed into the gold of inspiration. We live in a culture that encourages us to 'numb out' on all emotions. When we unfreeze, realizing with surprise the degree to which we were previously frozen, our awareness expands. We are no longer afraid of our emotions. We no longer label them bad or good or attach them to repetitive story. We feel vitally alive. Life goes from black and white to blazing beautiful color. There are no bad emotions. All emotions have their unique gifts to offer us. All emotions play significant roles and when freed can be felt deeply as they come and go, enhancing our experience of being alive. When we are no longer afraid of feeling our emotions – our main motivation for suppressing them – we can experience them fully and understand them as a profound source of power.

When is it beneficial to begin the treasure hunt for suppressed and buried emotion? To start with, observe your own life. What are your patterns? If you want to know where you are in your personal development, you need only look at the reality you currently inhabit.

How are your relationships, for example? If you have a significant other, is your relationship fun, supportive and loving? How are your friendships and family relationships? Your emotional health will deeply affect your relationships and we often find the same problems recur with new people. What we think is mad, passionate love may soon be revealed to be a marriage of the same frequencies of wounding when the honeymoon ends and the juggernaut with the baggage arrives.

How is your health? The wealth of information connecting our physical health to our emotional health is burgeoning.

Do you love what you do and where you live?

Do you have meaningful, purposeful work?

Do you see the universe as benevolent?

Do you feel empowered?

Do you feel you have a voice in your life and in the world?

Do you feel connected to your true essence, your inner source of power and abundance?

How is your inner-peace quotient?

Are you in a position to make contributions to the greater whole?

Are you part of a meaningful, nourishing community in which you can relax and be your true self? No masks? No anxiety?

What about your prosperity? Your happiness?

If you have identified aspects of your life that are not as you would like them to be, then you need to begin at the beginning by clearing your emotions. This attention to our emotional health is crucial and should be done by everyone. The final result is not only the huge benefits of unleashed power but also compassion and understanding for yourself and for others. When our power is available and we understand how to direct it, our dreams gain the fuel they need to manifest.

If you answered every question above with perfect satisfaction, then read on for the sake of others and find out how you may help your loved ones, your community and your work colleagues. You will also gain a greater understanding of your superpowers, your majesty and our incredible sentient universe. If you're in the middle on some questions, then I would still encourage you to read on and later go through all the exercises that appear in Part Two, giving them your best effort. Creative Alchemy techniques have helped many people of all ages and all walks of life heal and shift sticky emotional patterns, crippling emotional issues and even some physical ones while releasing enormous suppressed power, vitality and capacity for enjoyment, advancing human potential.

Humanity's Divorce from Emotion

How did we get ourselves in this pickle? The more civilized and industrialized humankind has become, the more we have favored the mind over the heart. Feelings have been downplayed as childish and irrational and reason made king. To call someone emotional is not usually a compliment and it brings to mind someone who is behaving irrationally.

It's possible the migration from a body-centric more holistic worldview to our current head-led reign of the cranial brain began as far back as the Neolithic agricultural revolution over 6000 years ago, when we began to impose our will on nature. [20] Where civilization has gone too far is in embracing reason above all else, placing an emphasis on empiricism (the theory that all knowledge must be proven by the physical senses). This opposes the development of *emotional intelligence, intuition* (the perception of that which is not visible) and the wisdom of the body, *gut instinct.*

Plato, the hugely influential classical Greek philosopher who lived around 2500 years ago, although a friend to the concept of divine inspiration, stated that which was apprehended by the mind was reason and that which was apprehended through the senses of the body was without reason. [21] In his *Dialogues* he explains our 'creation by the gods' as a sphere being fashioned, imitating the spherical nature of the universe, maintaining this sphere (our cranium) holds the soul and that it rests upon and is lord of a vehicle (the body) that serves as its means of locomotion. [22] That idea has stuck, influencing our development down the ages and reaching its lowest point with Richard Dawkins' metaphor of humans as fleshy robots – selfish survival machines. [23]

The European intellectual and philosophical movement called the Age of Reason (1685–1815) brought much positive advancement, including the concepts of individual liberty and religious tolerance. Its hallmarks were centricity of reason and empirical (observational) science. Again, emotions were very much downplayed. Religions have also played their part, dividing the soul from the body and casting the body, emotions and also women, in the role of the villain, distractions on the road to 'heaven.' Seventeenth-century French philosopher René Descartes famously stated, 'I think; therefore, I am.' This is actually very close to the truth, especially if we add our feelings to the quotient. However, he also created what is called the Cartesian split or mind–body dualism, stating the body cannot think, something which has definitely been disproven – more about that in the next chapter. Descartes saw the body as Plato did, something that could be dismantled and studied in its various physical components like a machine. Sir Isaac Newton, the 17th century classical scientist, one of the most influential of all time, believed only physical matter was real. The effect of the mind and the emotions on the body were considered psychosomatic, unscientific and not relevant. From the Stoics to the Renaissance, emotions were considered troublesome things, mistakes of judgment or psychosomatic phenomena to be controlled or discarded. [24] Despite our revolutions in technology and psychology it is clear these harmful ideas have persisted, slowing our evolution into a more capable and masterful species.

It's only since the mid-1990s that the traits of emotional intelligence have begun to be examined so that we can begin to understand their value to the individual and society. The field of neuroscience is confirming that restoring the connections between the mind, physical body, emotions and subtle energy body creates a potent collaboration. This adds intuition to intellect and increases motivation. It aids reason, creativity, empathy, connectivity, awareness, alertness and calm. It also heightens accurate perception and fuels manifestation. These revelations advance and accelerate our human potential in exciting new ways and in all fields of endeavor.

It is important to note that many indigenous cultures have avoided the compartmentalization of this predominantly European ethos, although it has crept into and permeated many cultures around the globe. Some indigenous cultures continue to use rituals and other methodology to connect to Earth wisdom and to the collective heart and mind of their communities for healing, wisdom and well-being and include emotional protocol in healing physical illness. In the middle part of the 20th century a revolution began. Many people turned to the practices of these cultures, desperately seeking wisdom their own cultures had lost. Now, various indigenous practices are being recognized by the mainstream for the enormous value they hold with many now forming parts of new wellness curricula. Scientists are also paying more attention to the wisdom of indigenous people who can still hear the Earth, the body and the cosmos, taking note of their observations and warnings and confirming their scientific merit. [25] China and India, as well as other cultures, never lost their understanding of the body as a network of energy systems, an understanding that has influenced the growing holistic and naturopathic movement in Western culture, which is increasingly confirmed by neuroscience and its offshoots, providing the basis for advanced quantum biological models currently emerging.

As the power of emotion becomes clear it has become more acceptable to state the obvious – the mind without managed emotions becomes a tyrant lacking in empathy, discernment and compassion, cut off from the wisdom and intuition of the heart and the pure instincts of the body. Wouldn't it be great if no one could govern or teach unless they had completed the process of Creative Alchemy and healed their emotions? Thankfully *emotional literacy* is a hot topic in education, parenting and business. Attitudes are beginning to change.

A Paradigm Shift

The teachings of ancient esoteric wisdom are being confirmed as modern science 'discovers' what the mystics always knew – the whole continually influences its parts and vice versa. There is no separation. This marriage of thought is creating a quantum leap in awareness.

The independently funded HeartMath Institute's studies on coherence between mind, body and heart and the influence that coherence has on our behavior, health and relationships has profound implications for this paradigm shift. Their research reveals the heart is a highly complex, self-organizing center for information processing that powerfully influences the cranial brain, communicating with it via the nervous system, pulse waves, hormones and electromagnetic fields, ultimately determining the quality of our lives. [26]

The heart was reclassified as part of the hormonal system in 1983. Some of the hormones the heart releases suppress emotional stress and increase motivation and behavior. It's also been discovered the heart produces and secretes oxytocin in amounts similar to the brain. Oxytocin is the love hormone that assists with childbirth and lactation but also social and pair bonding as well as tolerance, trust and friendship. [27] The heart is rightly considered the organ of love.

HeartMath data have revealed that the heart plays the leading role in coherence between our mind, emotions, body and energy field. Their studies demonstrate that self-generating a positive emotion in the heart such as affection, caring, or gratitude, while breathing deeply and evenly, causes the functions that process and allow the experience of physical stimuli to heighten and perceptual and mental processes to become greatly enhanced. Whereas uncontrolled emotion can cause mental chaos, managed emotion can heighten the efficacy of mental function. Being in a state of heart-brain coherence five minutes before a discrimination task produced a six-fold improvement in performance. [28] When many of us gather together in a state of coherence we have social coherence. A high level of social coherence is characterized by harmony, cooperation, trust, kindness and social stability, as well as a heightened organized flow of energy for efficient utilization.

Experimental research has shown that positive emotions, and the accompanying shift in our electromagnetic field caused by the resulting heart-brain coherence, can affect changes in our DNA. [29] Epigenetics, a branch of neuroscience, is a relatively new field that studies how the environment, which could mean intracellular, environmental, or energetic influences, impacts our DNA and biological systems. We once thought we were stuck with the hand we were dealt. It is now clear it is a movable feast. The implications are monumental. Further research has shown that the heart, which generates a much stronger electromagnetic field than the brain does, provides the energetic field that binds together the body's many systems as well as its DNA. [30]

Emotions hold enormous power over thought. Mental experience involves complex multidimensional mapping of physical and neural phenomena. [31] Emotions are so powerful that the processing of external stimuli and internal stimuli can be weighed equally in their effects. This means the emotions created by memory and recall are as affecting to our mind and body as experiencing these events in present time. *If you are constantly revisiting and repeating negative memories that addictively conjure the accompanying emotions those memories trigger, it is no different from exposing yourself to external traumatic events in the here and now over and over again.* You are reinjuring yourself. When you allow the catharsis and release of the original trauma, those inner tapes and resultant caustic chemical cascades will automatically cease.

Neuroscience has begun to chart how emotion-related input modulates cognitive input. How we feel greatly affects how we think. It is becoming clearer we need to raise our emotional intelligence to access the full potential of our intellect. There are more neural connections leading from the heart to the cognitive centers than vice versa! [32] It is now proven that our emotions affect our thinking more than the other way around and this supports one of the basic premises of Creative Alchemy – cleared emotions lead to clear thought.

The term 'cranial brain' has been coined to differentiate between the fairly recently named 'little brain' of our heart and the 'second brain' of our gut. Abundant supplies of neurons (nerve cells that form the basic working units of the brain) have been discovered in both areas. This sheds new light on the functions of intuition: *heart thinking* and gut instinct: *gut thinking*.[33] The little brain is also being credited with super-learning abilities confirming the ancient idea of heart-based wisdom. Sensory neurites have been discovered in other organs of the body and communication between these organs.[34] The parts are not separate from the whole. The mind is not localized in the brain. In fact, recent studies are beginning to confirm the mind and our consciousness also exist outside of the body, perhaps even emerging from a larger field of consciousness.[35] The brain functions as our biocomputer and can be reprogrammed. Magneto Encephalograph (MEG) technology can read our mental activity outside of our heads. Your thoughts are not contained in your brain![36]

Another leap forward in our understanding of emotion is that emotions operate more quickly than thought and can bypass linear reasoning entirely. Emotions can exist entirely independent of thought.[37] Emotional chaos causes mental health problems and mental stress. A clear, stable mind is dependent on mental and emotional congruency (being in sync) and proper management or mastery of the emotions. Really, mental health problems should be called emotional health problems. Emotional states activate the nervous system far more than mental states do. As neuroscience continues to examine the mind–body connection, it is clear that the boundaries between mind and emotion are blended. Emotional experience is a composite of brain stimuli and internal feedback of physical systems. What does this mean pragmatically for you? If you force yourself to think positively while your emotions are at odds with your goals you will increase the probability of unconscious self-sabotage. Suppressed emotion will affect the outcome of your manifestation every time creating discordance that scrambles your signals to the universal field.

The key finding that our thoughts have less power than our emotions is significant for the techniques of Creative Alchemy. We need emotion to fuel vision and the manifestation of our dreams. Emotions are our powerhouse. It's also why techniques such as positive affirmations and neuro-linguistic programming (NLP) are only partially successful if we don't address the healing of emotion. Discordant suppressed emotions will be sending out powerful signals that drown out any positive affirmation, although the use of affirmation while clearing emotion is very supportive.

We are now able to locate exactly where in the body certain emotions register. Researchers have charted which branch of the autonomic nervous system, either the sympathetic or the parasympathetic branch, responds most dramatically to individual emotions. They have also charted which organs are activated. For example, fear triggers the sympathetic nervous system and it discharges into the gastrointestinal and cardiovascular systems. Biologically, this is the reason why you might suddenly throw up or urinate or experience a pounding heart when you're shocked and frightened. This discovery has given rise to research which confirms that by mimicking the body's physiological reactions to emotion a specific feeling can be activated. This understanding is the foundation of the Perdekamp Emotional Method (PEM), which utilizes the knowledge of organ reactivity to emotion to reverse engineer the feeling process. This is an amazing method for actors who can learn to reliably create an authentic emotion night after night on stage, or take after take in a film studio. Using this technique, they can avoid falling prey to the stress caused by conjuring feelings through personal or manufactured memories and instead create authentic emotion using a primarily biological technique. Individuals with Asperger syndrome have had remarkable results using PEM, which can assist them in the recognition of different emotions in themselves and others. A PEM instructor helps an individual create emotional states by the inner physical manipulation of organs using breath, tension, energy, will and focused intention. This technique allows even the most intense emotions to be fully experienced in a safely managed way. [38]

It wasn't until the late 1990s that the brain imaging of traumatized Vietnam veterans vastly furthered our understanding of emotions and their physical corollaries. Brain-imaging tools such as functional magnetic resonance imaging (fMRI) began to show us what happens inside the brains of those who had experienced severe trauma, in this case from war. Bessel van der Kolk MD, one of the pioneers in the treatment of post-traumatic stress and visual brain mapping, observed that trauma detrimentally changes the way the mind and body manage perceptions and affects our cognitive processes. [39] Through this research, it was learned that trauma, large and small, deeply imprints our biology.

When our brains are in a state of 'fight or flight,' a common default position if we are triggered by buried trauma, our brain scans look like a hive of bees swarming. If we become frozen in fear and emotional paralysis, as many do when traumatized, the brain scans look like a blank slate. Areas of our brain that have to do with speech, reaction and communication literally shut down. Advanced brain imaging has given us a wealth of information, including visually demonstrating the profound efficacy of meditation for creating optimized brain states. Meditation balances the parts of the brain involved with external activities with the parts involved with emotions, two networks that are usually not active at the same time, creating a harmonious inner state. [40] These images have proved invaluable in our growing understanding of how to care for traumatized people and helped our understanding of the mind–body–emotion–energy connection. We are different systems of energy interacting either discordantly or harmoniously. Van der Kolk observed that trauma couldn't be entirely healed until we befriend the sensations in our bodies – our emotions, the messengers of our psyche. It is possible to befriend each emotion and understand its message for us with practice. But this takes courage, as we have been trained to avoid strong emotion at all cost. We are likely to eat, drink, shop, have meaningless sex, take drugs and otherwise anesthetize our emotions away in any way we can. We celebrate good news *and* bad news by anesthetizing ourselves. We are afraid of feeling. Time to shift the paradigm.

A quantum biological model of the body is emerging. Energy carries information. Obstructions in the energy fields caused by suppressed trauma can block the flow of information flow in our physical bodies and impede the innate ability to self-heal. In fact, we now know the body is also compressed energy. Ancient energy-based healing modalities coupled with modern medical and technological advances can create new holistic models of well-being. When we understand the physical role of electrical neurotransmission in thought and emotion it becomes easier to see how we are always interacting with and affecting the electromagnetic field that surrounds us.

Neuroscientists Alex Huth, Jack Gallant and their colleagues have created a quantitative model of the brain using fMRI to decode semantics from brain activity. From this research they are creating a word atlas of the human brain, displaying where individual words and the concepts they convey arise. [41] Though it's not perfected, we are now able to see electrical patterns of thought! These electrical patterns have power and frequency and are part of your manifestation armory and how you affect the world. No doubt there will soon come a time when we can completely correct old patterns through electrical stimulus. However, I believe the ability to consciously engineer our process step by step will lead to greater awareness as is more beneficial. This awareness can also assist with recognizing patterns in others and promoting compassion.

The 19th-century philosopher William James believed the greatest discovery of his generation was our ability to alter our lives by altering our minds. This relatively new understanding of our ability to alter our internal reality and therefore our external reality may be one of the most significant evolutionary factors accelerating our shift as a species from Homo sapiens to Homo illuminatus. It could be the most significant realization of our era.

The Messages of Our Emotions

Everything we feel matters. To be healed of trauma and suppressed emotion allows you to experience your emotions in a different way. Anger is our fire and the root of passion and compassion. It is also a natural response to our boundaries being transgressed. If we are frequently irritable it's a signal of suppressed anger. Grief signifies the measure of our passion and appreciation of life and is part of our human dignity. It is the price of love and all who experience it learn to respect it and become deeper because of it. But if grief becomes continual melancholy there's very likely an element of suppression involved. Shame or remorse can be warranted in certain situations but are very unhealthy emotions to become stuck in. Disgust can be a warning to step away, but if you feel it continually, it's a clue to a deeper story. Fear comes when we need to be warned of something. Unresolved fear becomes the base state of anxiety and even paranoia. Anxiety can accompany the suppression of any and all emotion.

When we allow ourselves to feel deeply the emotions considered 'negative' without blame or projection we find we can also experience the more pleasurable emotions more deeply. We feel more awe for life. Inspiration increases. Greater inspiration magnifies our ability to solve problems and innovate. The more we allow ourselves to experience our feelings deeply, the greater the harmony we will experience as well. We find we can maintain an even keel. Depending on our disposition, equanimity could feel like anything from peaceful contentment to bubbling joy. We are all unique.

Summing Up the Power of Emotion

The first steps of Creative Alchemy in Part Two will teach you many different ways to express and experience emotion. If you allow yourself to experience all your emotions fully, and express them in safe ways with the correct tools, they will flow through you and deepen your experience of life. It's only stuck or buried emotions that cause us to overreact, repeating thoughts like broken records, or suffer from anxiety and other afflictions. The first steps of Creative Alchemy use simple creative techniques, equally effective with adults and children, to access buried emotion. One reason these first steps are so effective is that the act of creativity represses the part of the brain that suppresses emotion. A bit of a twister. Creativity puts to sleep the part of our brain engaged 24/7 with ensuring you don't experience the disruptive catharsis of releasing buried emotion! Creativity opens the floodgates.

If you bring forth what is within you,
what is within you will save you.
If you do not bring forth what is within you,
what you do not bring forth will destroy you. [42]
– The Gospel of Thomas

You will begin to find your emotions are powerful allies. They expand our energy, our raw inner power and fuel our motivation. When we are in touch with our emotions, we've stopped living solely in the ivory tower of our heads and fallen into body wisdom with all its sensorial and intuitive gifts. Freed emotion is empowering.

Every time you are tempted to react in the same old way,
ask if you want to be a prisoner of the past or a pioneer of the future.
– Deepak Chopra

Our core emotional states create a bias that affects our belief system that in turn affects how we perceive the world. Two people could experience the same emotion, interpreting a cascade of chemicals differently. A person who feels safe in the world could interpret the cascade as the feeling of excitement. Another person, who does not feel safe in the world due to unhealed trauma, could interpret the same cascade as the feeling of fear.

If we have suppressed anger, we will tend to be defensive. Our attitude of defensiveness can cause others to react to us either defensively or offensively. Suppressed grief can give us the aura of the victim, attracting bullies and those who prey on the vulnerable, further enforcing our feelings of victimhood. Bullies are actually the reverse side of the same coin and often harbor feelings of low self-esteem.

Our emotional field extends farther than our mental field. Our energy fields exchange information with other energy fields, conveying our inner states to those we meet. The scientific principles of 'like attracts like' ensures we attract individuals and situations that match our energy fields. When we clear our suppressed emotion the information in our energy field also shifts becoming harmonious, attracting harmonious people and situations.

We cannot suppress selectively. When uncomfortable emotions are allowed to release we will be able to experience more joy, serenity and well-being. We are creating all the time. We have a choice to create consciously or unconsciously. We have a choice to use the substance of our life harmoniously or discordantly. If we have suppressed emotion creating discordance in our energy field, the patterns of manifestation are affected. We may self-sabotage or create a Frankenstein version of our dream. Harmony is the key to a steady flow of abundance and ease of manifestation.

- The energetic frequency of the emotions can be entrained and retrained by playing certain types of music, calming the emotions. Baroque composers were said to have especially understood this fact. Listening to their music has positive effects on the physical, mental and emotional frequencies and our whole energy field. This can assist in the weaning of receptor molecules addicted to stress hormones, encouraging them to draw hormones that imbue us with pleasurable feelings.

- Meditation is also crucial and can be as simple as finding a comfortable position with your back straight and your legs uncrossed and watching your breath. Bringing to mind something you are grateful for releases beneficial biological chemicals. Gratitude is crucial to any healing process and finding things to be grateful for daily helps achieve coherence.

- Walks in nature, time with friends and family, affection with pets, deep breathing, exercise, dance and creativity of *any* kind are very helpful.

- Keeping a journal is an excellent tool. It helps you to become aware of your patterns and triggers.

- When journaling, take some deep breaths and ask to be connected to your Presence. Ask a question and allow the answer to come from this profound aspect of your being which sees the bigger picture.

- Learning to acknowledge the emotions we are feeling and simply allowing ourselves to feel them without distraction, 'story' or blame is a powerful tool.

- Speak your truth without blame. Taking full responsibility for what you are feeling allows this. If you can honestly express how you are feeling in the moment without attacking, blaming or projecting, a deep alchemical process occurs.

The First Steps of the techniques of Creative Alchemy are dedicated to the all-important task of clearing our suppressed emotions.

2

OUR QUANTUM NAVIGATOR – THE POWER OF THE BODY

If anything is sacred the human body is sacred. [43]
– Walt Whitman, 'I Sing the Body Electric'

Our bodies are *quantum navigators*. Understanding the potential of your body is key to creating a radiant, grounded life and co-creating a paradise on our Earth. One of the characteristics of the shift to a new paradigm our planet is currently experiencing is a different attitude towards the body. This has occurred in various ways.

In the psychological and scientific community, advancements in the technology of emotion and in the sciences of epigenetics and neurology have really shifted perceptions of the body's purpose. We now know that the brain is flexible and we have 40,000 neurons in our hearts and 100–500 million neurons in our gut. [44]

New discoveries also spell the end of genetic determinism. Scientists have recently discovered that our thoughts, emotions and environments have such a profound impact on our DNA that you could 'inherit' your Aunt May's fear of spiders biologically, through chemical changes in the DNA. [45] This transgenerational inheritance is mediated by epigenetics. [46] Genetics refer to genes and gene function, epigenetics refers to gene regulation.

The widening acceptance of the mind–body–emotion interface has blurred the line where one ends and the other begins. Science is confirming that we are not brains in a flesh suit. Our bodies are highly intelligent contributors to our life experience and not just our means of locomotion.

In the spiritual and new thought communities, the current buzzword is *embodiment*. After decades of spiritual bypass with a trajectory that was about moving into the crown energy center and preparing to ascend, the zeitgeist is now about getting into the body, connecting to the root center, getting grounded, living from the heart and integrating our Presence, our Higher Self, into the body, creating a heaven of Earth. To treasure our bodies is to also treasure the Earth whose elements helped to form them. There's a growing understanding that loving our bodies is an environmental, spiritual and political act.

Being embodied allows us to be present, calmer and stronger emotionally and mentally. It gives us the power to manifest in the physical reality we inhabit and power to be a conscious creator of our shared world. When we are traumatized and emotional we shy away from the feelings stored deep within the body and tend to hang out in our minds, unaware of how limiting this perspective can be. Our bodies play a special role, functioning as extraordinary multidimensional vehicles with which we may navigate physical reality while sensing and interpreting nonphysical reality. Being embodied allows us to feel our emotions fully, 'read' our environments, decode information in the energy fields that comprise our local reality and experience pleasure. The electromagnetic fields of our four Earthly bodies, our mental, emotional, physical and energy bodies, have a potent effect on the fields surrounding us and can be affected and altered by them. Everything is communicating continually. Learning to be a master of energy is crucial to becoming a master of your destiny.

As quantum physicists develop more advanced tools to measure the electro-magnetic waves and particles of this universe, they have discovered that everything is composed of energy – and that energy is responsive and holds information. The material existence of a subatomic particle is called an excitation of the quantum field. It represents a quantum event wherein a wave of energy (nonmaterial potential) has collapsed into a particle – due to observation. From the research, we now know that the appearance of a subatomic particle is probable and not fixed. [47] Reality at the subatomic level is affected by attention. We can apply this discovery to support our understanding of how we affect the field by our attention. The perspective from which each of us views reality is just one subjective perspective in a multidimensional reality where countless potentialities and perspectives exist waiting for our focus. So, where should we place our attention? If we keep it on the energy, the quantum wave field, we can enter a reality where we can choose to be local and non-local, where we can communicate instantly over long distances, and where everything is fluid.

The bridge between quantum physics and human perception is the *consciousness observer moment* (COM). Just as quantum mechanics has shown us that the 'observer effect' collapses a wave form into a particle – a *quantum event* – a COM occurs when, out of the trillion possible realities that are awaiting manifestation in the vast information field of existence, our attention 'collapses' a particular reality into being. [48] You can focus your conscious observer moment on the reality you wish to call in and summon it. If you can conceive it and believe it, you can achieve it! This is a crucial element in becoming the master of destiny and physical reality.

With Creative Alchemy we are participating in a grand experiment of altering our reality, embracing the fact that we ourselves and everything in our world is composed of energy that we can receive information from the fields surrounding us and also influence it with our own energetic geometries of thought. The great mystics of old who studied the nature of reality believed that the energy that fills the Universe is conscious and self-aware. New science is certainly pointing in that direction and has gone so far as to call the field of energy that permeates and surrounds us information or data. Here's the leap I would like you to take with me. If matter is condensed energy and if energy is information, can we agree it is potentially consciousness? As I mentioned in the introduction, Max Planck, the Nobel Prize-winning founder of quantum physics, called the force that holds atoms in place and brings the atom to vibration *conscious and intelligent Spirit, which is the matrix of all matter.* [49] If we can accept that we emerge from this field of consciousness and information, we can begin to understand that we are actually made of consciousness itself. Our bodies are condensed consciousness able to apprehend greater consciousness while experiencing physicality. We can begin to look at our mind, body, emotions and energy systems as different frequencies of consciousness vibrating in synchronized harmony or discordance. When we are in harmony, the magic begins.

> *O I say these are not the parts*
> *and poems of the body only,*
> *but of the soul,*
> *O I say now these are the soul!* [50]
> – Walt Whitman, 'I Sing the Body Electric'

Our physicality is a great gift with its own wisdom to impart. The human body, including the brain, is an amazing sensing organism. It is the vehicle of our Earthly existence; our tool for perceptually apprehending realities that can be honed into an even more sensitive receptor with the ability to access the quantum knowledge of other informational fields. As Max Born, one of the early developers of quantum science pointed out, theoretical physics is a philosophy that has changed how we think of time and space, causality, substance, and matter and has taught us new methods of thinking that have applications far beyond physics. [51] It is now easier to understand that we are multidimensional beings in a sea of consciousness and the greater the degree of harmony we hold the further our personal fields of energy can reach. We become a being in perfect tune entraining all around us to greater harmony and well-being – Homo illuminatus, our Presence increasingly integrated into the expression of our physicality. The song of our cells – the keynote of our energy field – blends with the whole, uplifting all it contacts.

From the quantum information model of biology, we are learning that the human mind is non-local (meaning not fixed to the brain) and has the potential to merge with the infinite mind of the quantum field of energy from which matter arises. The human brain is a biocomputer built to access, interact with, and experience consciousness, as well as regulate physiological and psychological functions. It is a receptor, organizer, archivist and signaller. The sensing potential of the body can gather information from the field through the 'sixth sense' of intuition and 'gut instinct.' We are energy beings and even the quantum, leaping, whizzing probabilities otherwise known as our particles are energy. We can think of these bodies as energy fields of information interfacing with other energy fields of information through multiple dimensions.

Experiencing Biology as Biography

Working with the shaman and healer Peter Aziz in the West Country of England, a group of us made our bodies into batteries of heightened energy by eating only raw organic food for one month while ingesting Dead Sea salts. It sounds a bit weird but there was a method in the madness. It had the effect of greatly increasing the electrical activity in our bodies. We then gathered on retreat and Peter analyzed each of our mind–body matrixes with iridology (examination of the irises of the eyes). In the fractal, holographic nature of reality any aspect of the body holds the entire biography of the individual in various energy meridian systems and the eyes are especially revealing.

The first week released memories trapped in the cells of organs and flesh and the second week the skeletal system. Depending on what physical and emotional weaknesses were revealed in our iridology, certain acupressure points on the body were pressed to release a blocked emotional memory. The palate of the mouth holds the memory of conception, for instance. As emotional memory was released, extraordinary events occurred. For example, a man's curved spine straightened considerably. Those of us acting as assistants holding the points for others were often able to experience the event of another as it was released and even see the memory. For example, I and another facilitator began 'tripping' while holding on to the points of someone releasing psychedelics from his cells.

Most amazingly, a young man who had been hospitalized for schizophrenia released a collection of dark destructive thoughts that had taken on a life of their own, curing him. Just like in the movies, the room went very cold. He screamed a soul-shattering, piercing scream as the entity left. Peter 'caught' it and disposed of it. The person said he had a vision of being suicidal in a past life and the entity entering and staying for at least two lifetimes if not more. The young man had been in a mental institution and his mother had an amazing insight that there was something controlling him. Imagine how differently this story may have played out if his mother hadn't risked following her intuition and taken her son from a mental institution to a shaman, someone trained to negotiate the invisible realms.

Were the demons of myth and legend in fact destructive compulsive thoughts that began to cohere? Thoughts become things. Things are solidified consciousness. We are solidified consciousness. It's very possible and plausible. Most importantly, the person's life was changed. During emotional healing techniques, the body–mind matrix can reveal memories that are not in accordance with our current belief structures. In that case, we can think of them as metaphors.

The Language of the Body

On an energetic level, our bodies are tuning forks continually sending out a vibrational frequency. We are either sound as a bell or we are out of tune. Furthermore, we entrain others with our frequency, and they entrain us with theirs. To further explain this important word, *entrain* is a term used by acoustical and electrical engineers that describes one frequency causing another frequency to match it through synchronization. If you have a powerful resonant energy field, you affect the environment significantly and with a broad reach.

This vibrating sounding bell that is our being, when clear, resonates deeply, gaining access to increasingly greater elements of the whole. It moves with ease and grace, perceiving nonvisible realities through the visionary function of the pineal gland – the vision gland – and local realities through the other sensory functions of the body. As well as the famous five senses sight, sound, smell, taste, touch, we have senses that perceive many other things. These include pressure, temperature (thermoception), direction, time and more. Equilibrioception is our sense of balance within body movements – both their acceleration and directional changes. Different parts of our bodies, such as our taste buds, can move us to action because they contain chemoreceptors for perceiving toxic substances in the atmosphere around us and in our food. Nociceptors in our skin can perceive dangerous chemical and thermal stimulation so we can react abruptly to safeguard our well-being. We also have the ability to perceive magnetic fields, magnetoception. Our bodies are extraordinary.

If we do not allow ourselves to authentically experience and feel the free flow of our energy in motion – our emotions – we become dissociated from our bodies and all the sensory information that it has the ability to provide us. Our perceptions are dulled. When we are dissociated, we float above ourselves like helium balloons. The thought machine of the brain goes wild and we start to engage in repetitive, compulsive thinking (creating demons). We stew in our own chemicals under the stress of these thoughts as our adrenals pulse out adrenaline and other stress hormones that thrust us into a state of panic, leading to fight, flight or paralysis. We become anxious and fearful. We get ill. We are not privy to the wisdom of the body because we lose access to our gut instincts and our intuition – both sensorial and cognitive functions. We diminish our physical power and have no creative force, no charisma. Our boundaries are weak and easily transgressed causing anger. Internalized anger causes depression. We lose our power to move through the world or move the world. Our self-worth diminishes and we draw to us experiences that confirm our perceptions, the interpretations of which are rooted in old beliefs, spiralling downward. It can become dire if we don't find a way to interrupt the descent.

Your body is the book of your life. This densest aspect of your multidimensional being when clear of buried trauma and debris becomes an extraordinary sensing organ sending and receiving information. The heart sends signals to the brain that can influence perception, emotions and higher mental processes. [52] In fact, the heart sends more information to the brain than the brain sends to the heart. Our emotions emerge from the neurophysical matrix of our bodies. New technology has demonstrated that the heart's magnetic field is an information gatherer that extends several feet from the body. This field is at least 100 times stronger than the brain's magnetic field. The amplitude of the heart's electrical waves is 60 times greater than that of brainwaves, measured by electroencephalogram (EEG) devices. [53] It's important to note that current equipment doesn't measure beyond this, so it could be much greater. Recent studies confirm our ability to decode information in the magnetic field of another person. This is called cardio-electromagnetic communication. [54] This unspoken dialogue is an ongoing transfer of information.

As everything alive has an electromagnetic field, it makes sense this ability could be refined. As the heart carries more amplitude than the brain, it stands to reason those 40,000 neurons in the heart have a powerful role in decoding and transferring the information in the energy fields. This could explain how ancient indigenous cultures intuited the remedies inherent in various plants – by communicating with the plant's magnetic fields. They may have cultivated a heart-brain coherence that we have lost. Have you ever walked by someone and felt great sadness or elation? Or entered a room that felt creepy? Energy fields can be charged and leave imprints. Cardio-electromagnetic communication could be the mechanism of this perception. Our cardio-electromagnetic communication is further supported by mirror neurons. These recently discovered neurons react as strongly to watching an action as performing an action. Mirror neurons may form the basis of empathy. [55] Dysfunction of the mirror neuron system may explain conditions such as autism, a condition that can interfere with the interpretation of emotion. [56]

While the complex nervous system in the heart qualifies as the *little brain,* it is the enteric nervous system that helps to create the *second brain* of the gut. The small and large intestines combined are about 25 feet (7.5 meters) in length. Intestinal microbes help the intestines manufacture about 95 percent of our serotonin, the 'happy' hormone, which among other things contributes to our ability to sleep. [57] Scientists have been fascinated – perhaps even shocked – to discover that, just like the heart, there is more information going from the gut to the brain than the other way around. [58] The function of about 90 percent of the fibers in the vagus nerve, the main nerve between the brain and the gut, is to carry information from the gut to the brain. [59] This has literally turned our former understanding upside down. These recent discoveries have birthed the new fields of neurogastroenterology and neurogastropsychology. Mental and emotional health issues can emerge from imbalances in the second brain.

Understanding Your Magnificence

Scientists estimate that we have approximately 50 trillion cells in our bodies. Some have estimated as many as 100 trillion. You are the captain of this army of tiny sentient batteries on their own path of evolution. Each of these carry an electrical voltage equalling as much as a quadrillion volt inherent in our physical body! This is enough to light up a small city.

Do you know how an atomic explosion is created? By splitting atoms! Nuclear fission produces the atom bomb, unleashing incredible power released by the splitting of atomic nuclei. This power can destroy a whole city. The estimate for the number of atoms in a single cell is between ten trillion and 100 trillion. This means that we have around 7 octillion atoms in our bodies. That's 7,000,000,000,000,000,000,000,000,000. Think of the power inherent in your body. Imagine if that power was used for creation rather than destruction. You're a gazillion energy 'bombs' waiting to be directed towards the creation of something meaningful. They make up your organs, your skin and bones, your hair, your circulatory system, all of you. You have trillions of atoms at your beck and call. And here is something potentially mind-blowing to consider. Those atoms in your body are billions of years old. They were born in the bodies of stars blasted across space. On an elemental level, the human body is primarily composed of carbon, oxygen, nitrogen, and hydrogen. Every carbon and oxygen molecule on Earth is 7–12 billion years old. Most of the hydrogen on our planet was forged in the Big Bang, the current theory of how our universe emerged, an event now estimated to have occurred 13.7 billion years ago. As the Joni Mitchell song says: *We are stardust.*

The real kicker though is that 99.9999999 percent of your body is empty space. The atoms that make up the carbon, nitrogen, hydrogen, and oxygen molecules in your body (and everything else) are 99.9999999 percent empty. Well, not empty. They are energy fields. The same goes for your home, all of nature, everything!

If we removed the space, which is in fact energy, the particulate matter of the atoms of the entire human race would fit into the volume of a sugar cube. However, that sugar cube would have the density of a neutron star and weigh almost 500 million tons!

Coming to Our Senses

In the quantum era of integration and unleashed human potential, we are coming to our senses. By that I don't mean that we are becoming sensible. We've been 'sensible' for way too long. We are activating our senses so that they can rightly serve as the feelers of our reality. We are getting out of our heads and into our hearts and bodies. When we are embodied we feel the world and our empathy increases. We become averse to harming.

Our aim with Creative Alchemy is to stop thinking so much – to become empty headed but brilliant. You can become capable of accessing the highest fountain of wisdom accessible on a need-to-know basis directly from the quantum field, igniting knowledge with inspiration and finely honed intuition.

When you develop the courage and the patience and the caring to befriend and deeply feel the sensations of your body and experience your emotions as physical phenomena, a profound alchemy begins to take place. At this point, you have ignited a sacred fire that is gradually going to consume the self that is Homo sapiens and deliver you to the self that is Homo illuminatus. Ultimately, even the process of intending, visioning, willing, impassioning and manifesting will fall away. Evolving to greater states of expansion and deeper expressions of your life's purpose will become a *consequence of being*. Then you are truly in the flow. You are aligned and congruent. You've let go of the shore and the river takes you to new experiences and you enjoy the ride.

The Power of Congruence and Authenticity

Exploring energetic boundaries by working with horses, whose bio-magnetic energy fields and sensory capacity can stretch up to 60 feet (18 meters) outwards from their bodies, was a real game changer for me. I was on Sun Tui's farm in England with her herd of wild horses. Sun Tui is an equine facilitator who was one of the earliest trained to work with horses as facilitators for human personal development by the pioneering founders of Epona Equestrian Services, Linda Kohanov and Kathleen Barry Ingram. Her horses were unshod and ran free on her large farm. I was being instructed to halt the horses' progress towards me with my sheer embodied physical presence. The problem was I had none.

If we have been traumatized we often don't fully inhabit our bodies. At that juncture I was disembodied, or to use the medical term, disassociated, a reaction to trauma experienced in my youth. This gave me very weak boundaries. I had not developed the capacity to glean when or when not to allow the outer world into my personal space. When transgression to your personal space happens continually as a child you lose your discernment along with your self-worth. It was as if I had a sign on my forehead saying lunch has arrived. I was a walking invitation to bullies. When we are strong and healthy, we have good boundaries that say, 'This is my space, my body, mind, and emotions. My spirit. You are welcome if invited.'

I am also an empath, a person who has unusually heightened sensitivities and ability to read the emotions of others. My life at the time was an overstimulating deluge of emotions – both my own and those of others that I picked up. At that time, my consciousness was hovering above me, so afraid was I of the suppressed pain and chaos stored within my physical body, leaving me disempowered.

I was attempting to use my will to stop the horse by mentally projecting an image when Sun called out, 'Stop using your third eye!' Honestly, I felt like I was at Hogwarts School of Witchcraft and Wizardry. Sun could see and feel how disembodied I was. The horses knew it too, so they didn't respect my boundary and were walking right over to me, ignoring me when I said stop.

From working with the horses I realized that this was how I had been soldiering through my life, disembodied and using my will to affect reality with exhausting consequences.

During another exercise, Sun stood with me in the middle of the ring and asked me to move a horse around the edge of the ring with the force in my body. Huh? The horse just ignored me. Sun got in the center and just using her hips was able to get the large mare circling the corral to hug the fence and move faster or slower. It was amazing and a great illustration of the power we have when we are in our bodies. This 'being in the body' creates charisma. If no one listens to you, you are probably disembodied.

The horses at the farm were master teachers. They seemed to be able to know what needed dealing with – and for the women in my programme it was almost always our boundaries. Like many women in our culture and of my generation, we were never taught how to establish personal boundaries. In fact, we were taught to override our instincts, behave and serve without question, often leading to sticky and harmful repetitive relationships.

My work with the horses really helped me to understand what it meant to have a boundary and how to stand in my power. As Sun says, 'If you stand with all your apples in a basket in front of you, you can't blame the naughty ponies for coming along and snatching them. Tuck your apples away and give them out with carefully considered choice to those in your life who deserve them.'

Another important lesson from the horses was congruence. Their incredible limbic systems enable them to process a great deal of what you and I would consider hidden information. For example, if someone was covering up a fresh trauma – let's say they'd broken up with their partner that morning and came into the ring acting cheerful and chirpy despite a real state of angst – the horses would buck as if poked with a flaming cattle prod. If that same person came into the ring acting authentically vulnerable, the horses would surround her and nuzzle her. Horses have an amazing ability to root out incongruence.

Why does this matter? If we are suppressing emotion, we are incongruent. This affects the people with whom we are interacting and makes them somewhat wary of you even if they are not consciously interpreting the signals our energy field is projecting. You will also attract similar types of people and wonder why you are triggering each other.

Sun Tui's work is based on the understanding that stuck emotions leave us blocked and unable to move forwards in our lives in a meaningful and integrated way. She is a former soldier and master teacher in applied equine behavior and psychotraumatology. Her Taoist name translates to the 'art of living.' She has had great success with her Dare to Live programme, through which she helps veterans cope with post-traumatic stress disorder. Using the feedback of her horses, she helps them rewire pathways in their brains. For me, I became determined to be fully embodied so I could stay congruent and develop some presence and this work was an important step on my journey. [60]

Embrace what ever stage you are at with Radical Acceptance. Try to love every part of your journey. As long as you keep evolving there comes a point where it all makes sense. Practice staying in the body and being aware of the information it is giving you. This increases our authenticity. When we practice authenticity our words are steeped in truth and power. Next time a friend asks you how you are, tell the truth. Not a lengthy monologue and not a litany of woes and stories. Just the simple truth of how you feel. If you do it without self-pity, honoring the full spectrum of who you are, your friend's eyes won't glaze over. You'll be giving your friend permission to be honest with you as well. Your relationship will deepen. So many quietly suffer from shame that would disappear if they knew that for most of us, our bodies have been guesthouses for the same emotions.

The Guesthouse

This being human is a guesthouse.
Every morning a new arrival.
A joy, depression, a meanness,
some momentary awareness comes
as an unexpected visitor.
Welcome and entertain them all!
Even if they are a crowd of sorrows
who violently sweep your house
empty of its furniture,
still, treat each guest honorably.
He may be clearing you out
for some new delight.
The dark thought, the shame, the malice,
meet them at the door laughing,
and invite them in.
Be grateful for whoever comes,
because each has been sent
as a guide from beyond.

– Thirteenth-century Persian poet and Sufi mystic,
Jalal ad-Din Muhammad Rumi, translation by Coleman Barks.

How Trauma Silences Us

People who have experienced trauma often have difficulty expressing themselves. They can be driven into silence, as if their voices have been snatched.

What causes the silence? There are real physiological reasons why people
become silent after trauma. During trauma, a key language center in your
brain temporarily shuts down. It's called *Broca's area,* an area of neural matter
in the left hemisphere of your brain, behind your left temple and above your
left ear, that is involved in the expression of spoken and written language.
The hippocampus, an important part of the limbic brain, which is involved
in ordering emotions, memories and events, is also shut down by stress. It is
cortisol-sensitive and can atrophy from the continued onslaught of PTSD.

Since the limbic brain keeps us alive by interpreting connection, this part
of our brains is attuned to interdependence and the rhythms of relationships.
People enduring physical or psychological abuse can have a hard time holding
aspects of experience together because of the neurological harm caused by the

abuse that has weakened their ability to connect. Chronic stress can also impair the prefrontal cortex by disabling its capacity for ordered thought.

The amygdalae, two nut-shaped parts of the brain lying one in each hemisphere, are the body's fight or flight trigger; our guard dogs. The amygdalae are connected to our survival. In situations of chronic stress, they can start pumping out stress hormones to make the body ready to run from that proverbial wild tiger. Many of us are on continual alert. The amygdalae handle our memory, speech, and visual cues, providing us with our most-primal instincts, fear, hunger and arousal.

The Hans Christian Andersen story *The Little Mermaid*, about a mermaid who loses her voice to a sea witch in trade for a pair of legs, has always meant a great deal to me. I definitely experienced the stealing of my voice through trauma, unaware that there was actual physiological and neurological science to back up my experience. Like the Little Mermaid, I went through years as a silenced person, perpetuating my trauma and affecting my relationships.

Understanding the science of this phenomenon can also help us understand the shame and worthlessness so many abuse survivors feel. Working as a therapist, I would sense when I was treating a survivor of sexual abuse by the degree of shame they experienced. Why do people feel this shame? I think it is because we don't scream and because we are unable to scream, we feel somehow complicit. We suppress the emotion of the moment and lock it up tightly in a deeply buried inner box where it cries out to haunt us at inconvenient moments.

It has taken me a lifetime to get my voice back. I look back on a string of incidents where I was literally paralyzed and unable to act, sometimes impacting on those I loved dearly and I wish I had healed sooner. People who have experienced continual or extreme abuse commonly think that they are meant to just 'take it' and this thought lingers long after the original trauma. We need to rebuild our damaged self-worth if we have experienced trauma.

I don't think it's necessary to have experienced extreme trauma to have some of the same symptoms of chronic anxiety and low self-esteem because we live in a shut-down world where few of us experience unconditional love. So many are prey to feelings of abandonment and rejection because of this. We are all suffering to some degree and safely expressing our pain is necessary if we are to evolve and play an active part in creating our desired destiny and build a new world together.

My own healing journey brought me to many therapists, psychologists, sages, mystics, shamans, holy people, healers and holistic health practitioners around the world. I often trained in the modality I was exploring to better understand it and employ it. I feel honored to have learned from these masters who taught me of the inherent goodness of humanity above all and have added to my growing body of understanding of the Great Work, the path of alchemy by which the lead of the soul may be turned to gold.

My inner witness, which protected me from going over into the deep end, was so strong that I felt a little schizoid as I trained in these modalities and then applied them to healing myself and later when I had healed myself, in my work with others. In the end, the modalities I received from my Presence, which became the system that is Creative Alchemy, got me there. I am completely free of the intense emotional and mental trauma that haunted me for decades. Facilitating this modality and the further steps of empowerment it offers is my way of paying it forward, honoring the Law of the Circle.

Something good is happening in our culture. People are speaking out. Shame is dissolving in solidarity. People are uniting. Silence about abusive behavior is old paradigm. Tyranny's grip is eroding as people awaken to their true power and this happens most strongly when we stand shoulder to shoulder.

The Energy Centers of the Body

Although we have numerous energy centers (chakras) in our bodies, the seven main ones correspond to the seven main nerve ganglia that emanate from the spinal column. Each energy center has enormous power and corresponds to aspects of our human personalities, our qualities and gifts, our states of being and emotions, physical aspects of our bodies such as our endocrine system, nerve bundles and organs and even our planetary system. [61] These energy centers are the interface for our physical bodies and subtle energy field. The great masters of the East as well as the ascended master teaching assert they proceed from the greater quantum and become transducers for the manifestation of our physical reality, drawing in the subtle life force of universal cosmic energy (prana or chi) and transforming it into useable biological energy. Thought leaders with one foot in the world of human potential and the other in science such as cell biologist Bruce Lipton PhD, neuroscientist Dr Joe Dispenza and the late Candace Pert PhD, who launched the field of molecular neuropharmacology, [62] have given credence to the existence and importance of these centers, which have been acknowledged for millennia by ancient cultures. As we use the techniques of Creative Alchemy, our energy centers become clearer and stronger and we can consciously draw far more energy into our body.

The chakras were first mentioned in the Vedas (1500–500 BCE) as centers of energy throughout the body with countless rivers of energy flowing through them, noting that as well as the seven main ones there were countless more.

The Sanskrit word *chakra* means 'wheel' or 'sun.' Westernized interpretations emerged between the 13th and 16th centuries. It was the Theosophists in the 19th century who began to investigate the connections between the subtle energy fields and the biology of the body, particularly the endocrine system and nerve ganglions in the sympathetic nervous system. Noted psychologist Carl Jung and Anthroposophist Rudolf Steiner were among those who integrated Eastern spiritual concepts with psychological theories, referencing these energy centers in their work in the early part of the 20th century. More recently Anodea Judith PhD and others have built on those foundations leading to the system we embrace today. [63]

In the medical systems that have developed a quantum biological model of the human body, such as the traditional Indian and Chinese medicines, the wellness systems of many indigenous cultures and in the growing holistic systems of the West, the role of energy centers and pathways throughout the body called meridians, are taken very seriously. Thanks to advances in science and technology, all the energy phenomena in our bodies are measurable in incredible detail with new diagnostic tools, enabling doctors and scientists and health practitioners to measure the well-being of the body, mind and emotions via the state of the energy field. The development of special electrophotonic cameras that can capture images of electromagnetic fields means we can now photograph them, confirming their shape, color and alignment. These images can be used as diagnostic tools. These whorls of energy show us where they are in balance or out of balance, either by being too small and compacted, or spinning unevenly, and so on. No doubt we will be tested for certain jobs in the not so distant future by the rate of our vibrational frequency and the balance of our energy centers.

It's interesting to note that scientists and researchers have been attempting to capture this energy field in various forms of electrography as far back as 500 BCE in China. Were our forebears better able to see this electromagnetic discharge with their eyes, hence the very accurate religious paintings of electromagnetic energy fields emanating from heads and sometimes hands in medieval paintings known as halos?

As you eliminate stuck emotional debris your life energy will rise up your spine through these energy centers contributing to heightened levels of awareness and greater abilities including clairvoyance, the ability to see 'beyond the veil' or clairaudience, the ability to hear that which is not normally heard with our human ears. In both cases, it is an enhanced perceptual phenomenon caused by increased life force that enables us to see and hear that which is oscillating and vibrating at higher, faster frequencies. This phenomenon of *kundalini rising* is recognized in Eastern cultures that have advanced considerably beyond ours in their knowledge of the function of the energy centers and the energy meridians of the body. In the esoteric wisdom schools this energy is called *sacred fire*.

It is a natural effect of self-actualization but can be overwhelming if it happens spontaneously. Certain breathing practices can help ease the sensations associated with the spontaneous eruption of energy from the base of the spine as it pushes its way forwards and travels higher up through the body. These include pranayama and holotropic breathing. The same exercises can help activate kundalini to begin with. As you work in a focused way with your energy centers, you will find that stuck emotion relates to a particular area of the body and understanding the function of the relating energy center can help decode the nature of it.

It is possible to experience a healing crisis as the life force flows upwards, activating each center in turn, causing catharsis of anything stuck. For example, a crisis of the crown chakra might give you the feeling you must single-handedly save the world. When that center is healed, you will know what steps you need to take to be of service in balance with your life.

I can attest to the fact that the experience of the energy centers opening, healing and coming into balance can be physical and emotional. After applying the principles of Creative Alchemy for a while I began to have powerful experiences that I didn't fully understand. It was an experience of kundalini rising, or in the terminology of the ascended master teaching of the I AM Presence, sacred fire rising. Physicist David Bohm stated that we are frozen light. [64] When the sacred fire rises that light becomes less static and we experience the molten power of life itself coursing through us. Here's what I wrote in my journal at the time.

A Personal Experience of the Sacred Fire Rising

For the past few months I have been having these experiences that seem to me to be of essential God qualities. They last approximately three days. One was the experience of compassion. During this period, I was practically unable to speak or write for the whole of the time. I felt an unbearable sweetness, a breaking of the heart and a burning, almost painful expansion of the heart, a painful tenderness. Each time I was near someone I saw a literal bridge of energy between our hearts.

The first time these things started happening, I thought I had finally reached God consciousness and that it would be possible to live a life of bliss from then on. And then, when it went away, I felt abandoned as if I had been found and lost again.

Now, I am at the end of three days of feeling an enormous power. This is located in the upper abdomen in the area of my solar plexus. It feels on fire. I also have a strange feeling that I could live for a thousand years. It's a feeling of sheer power, of enormous, almost overwhelming life force.

Each time, the experience has crept up on me out of the corner of my eye and I have fanned it by focusing on it with an unbelievable result. This feeling of power is amazing. I am feeling unbelievably alive to the depth of my cells. Every part of me is electric and this electricity is contained, not projected. I stare into the sun and feel equal to it.

I have been told that this is the experience of kundalini rising which helps me not feel so bereft when it goes away. The color of the experience always relates to the particular chakra being expanded in the moment. The momentous feeling is the expansion. When I expand into it completely, it plateaus.

I asked before I went to sleep to be taken to the depths of my heart where the divine dwells, and all night long I was shown pictures of people who were not as they seemed at first. I was being shown how we can never judge another's path or choice. We are all learning what is right for us at that moment in time and the lesson is often very different than it first appears. The divine nature dwells in everything!

I saw a friend last night. I was still in the power experience and I became evangelical and very psychic with her. She has that effect on me anyway, but it was as if I expelled the mega-watt energy in me. After what became about two hours of a very intense hook-up with guidance and with realizations pouring through, I felt fatigued at which point my guidance said I must learn to keep the energy for myself as it will be working within me for the duration of the experience. The cup must be full and sustained before it can be shared.

Hearing this, I felt sad, as if I had blown it (literally).

The experience of power has gone but I am still feeling very strong and I am looking back upon my whole life with love. I am loving all the experiences and no longer feel pulled down or maligned by suffering. It is like a story I had read in a book now closed.

Rereading this description of my experience of 24 years ago, I am reminded to advise you to be very careful with your energy. When we start to feel massive spurts of inspiration it is our natural inclination to give it away. However, it is time to be selfish. When your cup is full to the brim and overflowing continually you will be ready to give energy away. In the meantime, you want to keep a large portion of your energy for yourself so you may heal all the cracks and fissures in your cup that drain you. This will allow you to continually fill up your energy field to overflowing. If you have children it is different, as we must always put our children first while they are small but teaching them boundaries is very healthy and doing your best to find time for yourself and solitude is very important, if possible.

Take some time to acquaint (or reacquaint) yourself with the meanings of the energy centers that interface our four earthly bodies. Each energy center rules specific functions, emotional states, organs and endocrine glands. The endocrine glands produce the hormones that are the rulers of our bodies. Understanding these main energy centers helps us practically understand how interconnected our four earthly bodies are. It's also very helpful if we have a physical illness and need to find the associated suppressed or out of balance emotion.

Advanced meditators can hear astral sounds particular to each energy center. The Self-Realization Fellowship founded by Paramahansa Yogananda teaches this technique beginning with learning to hear the universal sound of Aum. [65]

An Overview of the Functions of the Seven Main Energy Centers

Root Energy Center. *The Root,* located at the base of the spine, relates to survival, connectivity, groundedness and connection to Earth. When it is in balance we feel safe, connected, supported, strong, generous and trusting. When it is out of balance or shut down we can feel insecure, abandoned, angry, unsafe, selfish, obsessively worried about our basic needs such as money, food and shelter or fixated on past traumas. Being unrooted can be the cause of excessive mind chatter and jitteriness. It's crucial to begin working with the energy of this powerful center so that as your energy field expands you stay rooted, connected to the Earth and embodied. Many of us are cut off at the root, denying ourselves important sustenance, abandoning ourselves. We float above.

The root chakra is associated with the adrenal cortex glands that regulate the immune system and the metabolism. Organs involved with this energy center are the testes, kidneys and spine. It emanates a red color that can be picked up by photonic-sensitive cameras.

Check the root of your body. Feel for the clenching of muscles. Soften and relax. Pull your breath through your body to the root. Feel your sit bones expand and your perineum and anus muscles relax. This is also beneficial for the prostate gland. If this area is tight we are 'uptight.' Put your hand on your lower abdomen as you breathe and feel it balloon out. Babies do this naturally. Breathing in this way can shift you from the sympathetic nervous system into the calm cool parasympathetic in moments. The astral sound is like the buzz of the bumblebee.

Sacral Energy Center. *The Sacral Energy Center* is based below the navel. It relates to creativity, sexual energy and relationships. When it is balanced, we give and receive in balance. Feelings of bravery, passion, confidence and excitement flow easily. We feel social and connected to others. When out of balance we can feel unsure of ourselves in relationships, fearful, anxious, vulnerable, isolated, self-pitying and morose. We can become withdrawn or chatter nervously. Creativity helps balance this energy center and healthy loving sexuality.

The sacral chakra is associated with the reproductive glands, which control sexual development and secrete sex hormones. Organs involved are the urinary bladder, ovaries, testicles, kidneys, gall bladder, large intestine and spleen. It emanates peachy orange.

Bring your breath into this energy center. Use the power of your breath, mind and imagination to free the flow. Sense the vibrant peach color. The astral sound is the sound of the flute.

Solar Plexus Energy Center. *The Solar Plexus* is located in the abdomen. It relates to will, gut instinct and personal power. It's the seat of our personality. When it is in balance we are like a sun radiating positive self-esteem, self-respect, compassion and assertiveness. We feel strong and full of life. When out of balance we can feel powerless, worthless, self-critical, ashamed and self-conscious, or a need to dominate and control from fear and lack of trust in the natural flow. Good healthy food and exercise help balance this center.

The solar plexus chakra is associated with the pancreas, which regulates metabolism. An out-of-balance pancreas can fuel a death wish. Other organs involved are the small intestine, liver, stomach and upper spine. It emanates yellow. Breathe deeply into this center. Visualize it emanating warm rays that flood your body, mind and emotions with a radiant golden-yellow glow. The astral sound is a harp!

Heart Energy Center. *The Heart Energy Center* is located in the area of the physical heart. It relates to loving feelings, empathy, unconditional love and affection. Unconditional love is the absence of judgment plus compassion. This energy center relates to our humanity and general well-being. When it's in balance we feel joy, gratitude, trust and forgiveness. When out of balance or shut down we can feel jealousy, anger, bitterness, rejection and flightiness.

In the womb, our hands grow straight out of the heart. The first steps of Creative Alchemy are very helpful for this energy center, as is any creative act. Getting your hands in the soil, doing some gardening and having a practice of service in your life are also useful.

The heart chakra is associated with the thymus gland, which regulates the immune system. Organs involved include the heart and lungs. It emanates green.

Most of us have felt the opening and a burning sensation when we are in a state of deep appreciation and gratitude, love or caring. Breathe deeply into the heart area. Breathe in and out and picture your breath forming a wind flowing through it, clearing all the pebbles. So many of us are closed here and not even aware of it, greatly diminishing our experiences of the world. It's calming to bathe this energy center with soft pink light, vocalize from the heart and conjure up memories of times when you were happy, caring and appreciative. Bring to mind someone or some place that you love. When we connect with our heart in this way, we trigger heart-brain coherence and all its numerous benefits. Regarding the pink light, remember your thoughts are formed in electrical fields. EEG (electroencephalogram) technology shows that mental activity can be effectively analyzed and monitored entirely by electromagnetic principles. The electromagnetic frequencies of colors impart unique qualities that relate to their specific oscillations and vibrations. The astral sound of the heart is a bell-like or gong sound.

Throat Energy Center. *The Throat Energy Center* is located in the physical throat. It relates to speaking your truth, your voice in the world, expressing your essence, communicating your needs and taking responsibility for decision-making. It is strongly connected to the truthful expression of our unique essence.

When in balance we can communicate easily and freely. When out of balance we feel powerless to speak out, afraid of judgment and rejection, sad, hurt, disempowered and resentful.

The throat chakra is associated with the thyroid gland, which regulates body temperature and metabolism. Organs involved include the bronchial tubes, oesophagus, vocal folds, mouth and tongue. Singing is wonderful for the throat chakra and humming. Wild lyrical non-verbal outpourings from the heart are very beneficial, as is a good roar. It emanates blue. The astral sound is like the roar of a distant sea.

The Third Eye Energy Center. *The Third Eye* is located between the eyebrows. It relates to the imagination, dreams, clairvoyance, Living Vision, spiritual wisdom, understanding and intuition. When it is in balance we can differentiate between truth and illusion. We're clear, empathic, focused and open to receiving wisdom and insight. We can perceive beyond the physical. When it is out of balance we are overly critical, sceptical and unable to self-reflect. Life is black and white.

The third eye is associated with the pituitary gland which governs the other glands, and the pineal gland which regulates sleep and circadian rhythms as well as the reproductive hormones and is considered the 'gland of vision.' Organs associated include the eyes, nerves and brain. It emanates indigo.

Breathe into the third eye and bathe it in appreciation. Don't wear sunglasses unless driving or if you do, take them off occasionally. Sunlight and love are stimulants for the pineal gland. Daydream. Build positive pictures in your mind. Look at or make art. Listen to beautiful classical music and imagine scenarios. We'll work a lot with the imagination later on and build strength here to create the Living Vision needed for conscious creative manifestation. The astral sound is a cosmic symphony of vibratory sound.

Crown Energy Center. *The Crown* is located at the top of the head. It relates to connection to the universe, dynamic thought, truth, wisdom, unity consciousness, integration, living in the present, feeling life has meaning and devotion. It is where we perceive the sacred and where we experience gnosis, knowing beyond the intellect.

When it is in balance we feel expansive, empowered, calm, valued, trusting the flow of life and can easily access inspiration and insight. When it is very activated we can feel we are divine, one with our Source and true essence beyond the limiting constructs of the earthly vehicles. We feel committed to a purpose beyond ourselves. When out of balance or shut down we can feel confusion, alienation and depression. We may develop rigid religiosity or 'us and them' spirituality and feel very constricted. We can also feel we need to save the world.

Organs involved include the spinal cord, nerves and brain stem. The best tonic for this energy center is meditation, stillness, prayer and contemplation. Devotional feelings and gratitude towards life open it. It emanates violet-white.

Other important energy centers are in our hands and feet and at the top of the spine. We also have energy centers outside of the physical body in our electromagnetic energy field, some of which assist us in creating deeper connections to Earth, transcendence of time and deeper galactic connections and perceptions. These transcendent energy centers are not connected to human emotion or associated with endocrine glands. In every multidimensional aspect of ourselves, we will have subtler and subtler energy centers.

There are current schools of thought that our carbon-based bodies are becoming silicon based and will continue to evolve into solar crystalline bodies. As the vibratory rate increases, the space–time continuum will shift and we will become fifth dimensional beings as third dimensional density expands and is finally transmuted. The solar crystalline body has 12 main energy centers called the *solar crystalline aspects of deity*. [66] When one of these is called into action by decrees to our Presence, all 12 temporarily radiate the same color and qualities creating an enormous focus and battery of power. For example, if we call into action the fifth solar crystalline energy center of illumined truth, inner vision, healing, concentration and consecration, all 12 will emanate like emerald-green suns, projecting the frequency of that quality into the surrounding energy fields.

It's possible when we're in heightened states of awareness to see the colors emanating from the chakras with the naked eye. A mother lovingly speaking of her children may flow green from her hands. An angrily shouting person may emanate red from his head which is the reverse of where it should be – he is literally becoming a hot-head.

The Vagus Nerve

We've touched upon the energy centers in the body and the multidimensional, quantum nature of the human energy field as it intersects with the unified field. We now know of the intimacy of the mind–body–emotion–energy connection.

There is a part of the body it is crucial to highlight and that is the vagus nerve. This nerve connects the brain to the heart, lungs, stomach, spleen and every organ in the body except the adrenal glands. It even connects to the muscles of the larynx that allow speech. It's a crucial regulatory component of the gut-brain axis.

The vagus nerve, which is the tenth cranial nerve, starts above the midbrain and descends to the colon, controlling eye movements, facial expression, heart rate variability, tone of voice, and breathing and the diaphragm and energizes all the organs from the brain down through the intestines to the colon. Bidirectional, the signals that it carries move back and forth from brain to body and body to brain. *It is thought to be essential for the distribution of life force.*

The vagus nerve is the most important nerve in the autonomic nervous system. As such, in ancient wisdom teachings from India it is referred to as the kundalini 'snake.' It is said that kundalini coils three times as it travels up the spine, just as the vagus nerve connects three times to the spine. It also continually communicates with the enteric nervous system (ENS) that is embedded throughout the whole gastrointestinal tract, where 100–500 million 'second brain' cells are located, the largest accumulation of brain cells in the body outside the cranial brain. [67]

The autonomic nervous system has three main branches. The parasympathetic branch is responsible for the relaxation response. The sympathetic branch is what snaps into action when our fight or flight response is triggered. The ENS modulates blood flow and endocrine and immune functions, among other things. The vagus nerve is the main component of the parasympathetic nervous system. Stimulating the vagus nerve creates relaxation and soothes the sympathetic nervous overdrive which so many of us are in 24/7. This overdrive causes inflammation of our tissues, is taxing on our internal chemistry and can lead to depression and illness.

When the vagus nerve is stimulated, it releases beneficial enzymes designed to improve immune system responses, triggering the body's healing resources with the potential to alleviate physical and emotional illness, including major depressive disorders. Studies have shown that self-generated positive emotions greatly increase vagal tone, as does yoga, qi gong and other practices. Yoga breathing (*pranayama*) improves mood, stress levels, cognitive functioning and autonomic nervous system regulation. Anger and depression suppress the vagus nerve. As you clear suppressed emotion from your body your life force energy will distribute itself more evenly throughout your energy centers and the corresponding physical organs.

A few of the many ways you can stimulate this important nerve are exercise, smiling and taking long slow breaths, splashing with cold water and meditation. Results on the health benefits of meditation from research funded by the David Lynch Foundation, which was founded with the mission to teach Transcendental Meditation (TM) in schools and with at-risk populations, show that meditation makes us happier, raises our immune response, reduces post-traumatic stress, increases empathy and compassion, improves learning ability and recall, increases confidence and creativity, reduces symptoms of attention deficit hyperactivity disorder (ADHD), reduces anxiety, stress, depression, violent conflict, helps with recovery from abuse and much more.

Simple meditative practices, like following the breath or breathing long slow breaths while adding a feeling of gratitude or appreciation, will change our brainwaves and release beneficial hormones, enzymes, and other positive neurochemicals that enhance every area of our lives. It's also one of the main ways to receive consistent guidance from the Presence. Brain imaging done with Tibetan monks while they were meditating shows that the part of the brain that locates us in space–time turns off, allowing for the feeling of interconnectedness we can experience while meditating. We should all meditate, whether it is a simple breath meditation, a mantra meditation, or any practice which brings stillness and coherence to your being. You will be stimulating your vagus nerve, releasing a host of feel-good hormones, as well as potentially relieving addictive behavior.

My practice is the Kriya Yoga meditation that Paramahansa Yogananda brought to the West and called the *airplane method to reach God*. He stated it was the real fire rite extolled in the Bhagavad Gita, the alchemical flame that transmutes the dross and madness to purity by bathing the body in undecaying light, defying illness and even death. [68] Paramahansa Yogananda achieved the pinnacle of mastery. He left the world in full consciousness (called Mahasamadhi) [69] having informed his followers earlier he would be leaving the body. He did this in front of a large audience after a beautiful speech regarding his love for 'Mother India,' raising his eyes upwards and consciously leaving his body. He was cremated 20 days later with no signs of corruption or decay. [70] Kriya Yoga, or Royal Yoga is taught by the organization he founded, Self-Realization Fellowship, if it appeals to you.

The vagus nerve even influences our genetic profiles. Keeping it healthy is an important tool in our Creative Alchemy practice.

Simple Ways to Accelerate Your Quantum Navigator

Every morning, pause and thank every bit of your body right down to its electrons! Thank your organs (brain, heart, kidneys, liver, pancreas, gall bladder, skin, stomach, intestines, your glands, your circulatory, digestive and nervous systems and so on). Call your receptor molecules to attention and reset them to attract positive neurotransmitters such as serotonin that increase feelings of self-worth and happiness and decrease anxiety and depression. Send love to all your cells and every system in your body. Call your team into peaceful and beneficial collaboration.

In the next chapter, you will learn how Professor Masaru Emoto's water molecule tests showed us the impact of thought on water. Every thought and emotion carries an energetic signature and his studies showed that the word *hate* or *love* was enough to negatively or positively transform the shape of the water molecule. As we are approximately 70 percent water imagine the power of strong emotive thought! Sending love and gratitude to your body shifts the signature of your energy field into harmony and is deeply beneficial for the regulation and harmony of all systems.

We will learn how to call in the energetic signature of an element or remedy. There will come a time when we will have developed such advanced technology of mind that we need only to adjust our frequency to effect cures. Many forms of energy medicine and energy technology are already being employed that can help us attain such results. Loving yourself is sound energy medicine. Love is a powerful magnet for all good things. When pure love is the cause, the effect we draw is also love.

Also, thank your emotions and your thoughts. They are all looking to you for guidance. Before you get out of bed, connect to your Presence and give thanks for your life. You will be given a powerful technique for this in Part Two. Imagine the sort of day you will have and send out rivers of goodwill. Lay down a carpet of benevolence for you to walk on and to enfold you and all you meet.

Ask your immune system to please operate perfectly, your blood pressure to calibrate to optimum, your stomach, oesophagus and intestines to digest well. Develop a relationship with these functions filling them with the heightened emotion of love and caring. Tell yourself *you love you*. Breathe into each of your energy centers, blowing out any debris into a Violet Fire. The alchemical properties of the violet spectrum will be explained in Chapter Five. Just as soap and water cleanses our body there are energy techniques to cleanse our energy field that are very powerful. Put on some music and move your body. Stretch. Dance! Sing!

Ask your emotions to be expressive and find equilibrium. Ask your mind to be illuminated and to concentrate on being brilliant instead of stopping you. Ask for helpful friends and perfect synchronicity and wild opportunity. And as you do, *pause, see it, feel it.* Suffuse your body with the positive emotion of delight that confirms you have already received it. Collapse that potentiality into the quantum event of a manifested particular reality!

Every part of you is alive, conscious, and responsive. Your 50 trillion-plus cells are waiting for your loving appreciation and gentle command. Be imaginative.

Summing up the Power of the Body

Our cells emit a frequency which informs the frequency of the bio-magnetic aura or energy field that surrounds and permeates us. This energy field is communicating with other energy fields at all times. The scientific law 'like attracts like' is always in operation. We will draw situations to us that mirror our frequency. If we have low self-worth, self-pity, patterns of suppressed grief or anger, our energy fields will be busy sending signals that draw situations to confirm and strengthen these. If we love and respect ourselves, and have done the work to clear our emotional and mental bodies, we emit harmonious frequencies, which draw harmonious situations.

If we are detached from our physical body and ignore it, it can't do its job. We are unable to take advantage of its phenomenal sensing capabilities such as the intuition of the little brain and the gut instinct of the second brain, severely hampering our navigational abilities as we journey through life. We'll be driving with the headlights off. If our body's sensorial capabilities are allowed to operate optimally, we will gain insight, pre-cognition and synchronicity, as well as much stronger discernment regarding what is right for us, where we should focus our attention, what we should eat and so on.

If we are cut off from our bodies and stuck in our heads there will be no force or presence in our body. We will have a hard time affecting our reality. We may be overlooked and ignored or preyed upon. If we are embodied we have strong boundaries, presence, charisma and force in the world. Our quantum navigator is alive with electricity and abilities to sense and interpret the world around us.

Put on some music and dance wildly. Jump, which is also good for the lymphatic system. Breathe! Pat yourself. Hug yourself. Ask your body what it needs and *listen*. We tend to continually override the whispers of our body. How many times have you been deeply intent on a task and 'heard' *I need water* or *I need some time in the park* or *I need a break* and just kept going? How many times have you 'heard' your body whisper *don't drink that, don't have another one, don't eat that, get some sleep now*, and completely ignored it? Or how about, *don't go off with that person? Don't enter that room*. Or *do take that walk* and *do enter that bookstore* and *do purchase that book* that practically fell into your hand. This is all body wisdom.

The body is equipped with extraordinary sensing abilities which work in tandem with your other three earthly bodies, your mental, emotional and energy body and your cosmic bodies, the Causal Body, the Presence and the bridge between the two, your Higher Mind. The more you honor it and the more you learn to listen to it the more exciting your journey will become.

1. For anxiety, warm baths followed by self-massage with a good quality oil such as organic olive oil or sesame oil which are also nutrient-filled. (Your body absorbs 60 percent of what you put on it, always try for non-chemical and organic.)

2. Hold your body, stroke it and tell it you love it, and that everything is all right. Be its loving parent and guardian. We often treat our pets better than our own bodies.

3. Listen to it. Quieten down, take some breaths and let your body have its say. Does it need water? A walk? Some stretching? Go for a run? It may have some very surprising advice if you take the time to listen.

4. Lie on the Earth. Our bodies adore being in nature and it refuels them. Walk by the sea, lie on some grass or take a walk in a forest, whatever you can do.

5. Take frequent breaks from work, especially if work involves computers. Splash cold water on your face. Think of something that makes you smile.

6. Thank your body every day.

7. Dress it up in nice clothes!

8. Meditate!

9. Slow down. Lovingly prepare and consume your food and get the freshest, highest quality you can. Your body's worth it and will love you back big time if you do.

10. Breathe into areas of your body, organs, energy centers and so on, and write down messages in your journal.

Once we've cleared our emotions and begun to lay down new neural pathways and new patterns of behavior for our bodies, life gets easier and easier. There are Creative Alchemy techniques in Part Two to help you hear what your body is saying and begin a meaningful dialogue which are very effective. I have witnessed and personally experienced these techniques creating access to body wisdom resulting in very in-depth direction and advice which has been astounding to myself and others. This is the journey Creative Alchemy will support you through.

3

OUR QUANTUM CONDUCTOR –
THE POWER OF SOUND AND ENERGY

God geometrises continually. [71]

– Plato

Our Universe is comprised of energy waves forming geometric patterns of harmonic frequency. The energy waves oscillate (expand) and vibrate (contract), creating frequency. We can influence these harmonic patterns and affect their shape and also be influenced by the harmonic patterns around us. We can speak of these patterns as architectural templates or musical scores: both have mathematical foundations. Our Universe, from the smallest daisy to the largest star, is an infinite interweaving of these energy patterns, these templates. We live in a symphony of audible and inaudible sound.

An example of our mathematical universe is the Fibonacci sequence, a sequence of numbers that is created by continually adding the last two digits of the preceding number. This sequence geometrically becomes the Golden Ratio, a shape of perfect proportion best expressed by the Nautilus Shell and the work of great artists who use the ratio to create perfection, such as Leonardo da Vinci. The numerical sequence reappears continually in nature, in the spiral of sunflower seeds, the fruitlets of the pineapple, the woody petals of the pine cone, the curl of a wave, the petals of flowers, the branches of trees, the measurements of the human body, DNA and so on. [72] This physical appearance of harmonic sequencing, bearing the imprint of the musical frequency that creates it, could be called visible sound.

The 17th-century physicist, philosopher, mathematician and astronomer, Galileo Galilei, was one of the first scientists we know of to observe that the relationships of astronomical objects in our universe could be described mathematically. He believed all of nature, the entire universe, was a great book written in the language of mathematics. [73] Galileo was not the first scientist to go on record with this theory. The Pythagoreans of 5th-century BCE Greece married mathematics and music in their description of the cosmos, coining the phrase *music of the spheres*. Having observed the ratios of whole notes in music, Pythagoras hypothesized that the same 'harmonies' would exist between planets. Contemporary quantum physicists working to develop superstring theory may have been inspired by the same analogy. They have observed that standing waves of electrons propagating through space are harmonic and their behavior can be mathematically rendered. [74]

Massachusetts Institute of Technology professor Max Tegmark also argues that the Universe is a giant mathematical object. He explains how it is composed of geometric shapes and that all the patterns in nature can be described mathematically, from motion and gravity to subatomic particles. [75]

Nobel Prize-winning physicist Erwin Schrödinger PhD described the wave oscillations of electrons as spherical harmonics. Physicist Louis de Broglie noted that the oscillation patterns of electrons will correspond to frequencies and levels of energy just as a plucked guitar string would. [76] It was de Broglie who first postulated that if light was made of discrete particles (now called photons), then matter might have a wave nature. [77]

What does all this have to do with Creative Alchemy? Understanding that we are frequency beings creating patterns that impact upon form is crucial to understanding how we affect and interact with our environment, and also how we may heal ourselves and accelerate our potential. Energetic wave patterns made by individuals or groups in a state of internal harmony engage with the quantum energy field in a way that can positively affect events and outcomes, and the reverse is also true. Discordance creates discordant manifestations and harmony creates sublime ones. Everything has a frequency. This frequency with its oscillations and vibrations creates a subtle music we respond to or recoil from. Healthy sun-grown foods emit very high frequencies. Junk food is lifeless, emits zero frequency and robs us of life force as we digest it. A healthy body emits a high frequency; a sick body emits a low frequency. A dead body emits no frequency. It's not fanciful, *it's physics.*

Our subtle energy body, which we will call our *quantum conductor*, is communicating with the world around us on a continual basis. This energy body, also called the etheric body or dream body, holds the memory of every thought, word, emotion and act we have created in all timelines and dimensions. Clearing this record of discordant memory so that you may experience 'fresh perception' and be free of the pressures of this accumulation is one of the goals of Creative Alchemy. Our planet's subtle energy body also holds the memories of all its inhabitants and its own journey through time. These powerful patterns can impact upon life and make change more difficult. Some of the techniques of Creative Alchemy also address cleansing this planetary memory bank for the sake of humanity's evolution and all life on Earth.

Our energy body is constantly communicating with its surroundings. When it is clear of debris it can play its role as the conductor of the fields surrounding us and insure our physical health and vitality. Illness is likely to appear as a blockage in the energy field before manifesting in the physical body. The energy body is charged with holding our template for physical incarnation and is also the one we travel in at night. Some of the dreams are actually awareness of these travels. As we clear the memory body by releasing the suppressed and unhealed emotion this aspect of ourselves can begin to do its job and we may notice we can add 'growing younger' to our list of benefits.

We each have a keynote that holds the information of our true essence and our purpose. The clearer we are the more exquisite and resonant our harmonic expression becomes, finally enabling us to sing our true song and take our rightful place in the symphony of life. As we become clearer our frequency quickens. This quickening causes the expansion of consciousness. In the practice of Creative Alchemy, we make use of the power of sound for the purpose of raising our consciousness and helping to dissolve stuck emotions from our bodies.

By repeating certain sounds aloud or internally that have very high frequencies you can reverse-engineer your current state of consciousness to the states of expanded consciousness the sounds represent, thereby altering your frequency. This is the science behind repeating affirmations and decrees. They must be repeated with dedication, emotion and consistency, gradually replacing our negative thought patterns of judgment, recrimination, doubt and regret with high vibrational qualities of confidence, self-love, compassion and victory.

Your energy field becomes in tune and as clear as a bell. You reach a state of harmony and have consistent inner peace punctuated by moments of joy, an inkling of your true nature. Very importantly, you feel safe. You have become your own ideal parent and you are connected to the Source of your life and abundance. Anxiety is for the most part gone, as is any sense of competition or comparison, for you know your job description is to become your own glorious unique authentic self and there is only one of you!

Hospitals of the future will be filled with chambers of light, sound and color that will entrain the bioenergy field of the individual to a higher frequency, healing illness. Illness will be discovered to be a result of discordance in the mental, physical, emotional or energy bodies. The use of laser for medicinal purposes, already in use, will replace surgical tools. Discordant music and druggy events will be replaced with pleasure palaces of light and color where connection to the Presence, the source of exquisite natural states of bliss, can be achieved. People will dance in temples of sound to rhythmic, flowing music, expressing various states of being to recharge and inspire.

We are energy beings. Waveforms. Beings of sound and color and light. The mind is an electromagnetic field made of trillions of photons of light interfacing with the information in all electromagnetic fields. Particles of light called biophotons control the function of the entire human metabolism, transmitting information between and within cells. New discoveries show that genetic cellular damage can be healed by quanta of light. [78] Our well-being correlates to our vibrational frequency.

Frequency is measured by the number of waves or cycles that a sound makes in a second. We use Hertz (Hz) to measure cycles per second. Our brain states can be measured in frequency and the brainwaves of the gamma state, the expansive state associated with unity consciousness, has the highest and fastest Hz, lending credibility to the idea that consciousness can be measured. We are continually creating coherence and congruence or discordance (lack of harmony) and dissonance (inconsistency between one's beliefs and actions), which affect the energy fields our physical, mental, emotional and subtle energy bodies emit. In this way we entrain our surroundings like tuning forks and are also entrained by our surroundings. The techniques of Creative Alchemy are designed to strengthen and empower your energy field so that you withstand the assaults of the fields around you and positively affect them, restoring them to their own intrinsic harmony.

Let's revisit these important terms. In relation to humanity, the term *congruence* is when feelings, actions, experience and awareness are aligned. [79] *Coherence* is the consistent, integrated, connected, efficient relationship between the whole and its parts – harmony. These measurements have become a key concept in quantum physics, sociology, brain and consciousness research, cosmology, physiology and among people interested in global and planetary well-being. [80] These are the ideal states of being that Creative Alchemy aspires to lead you to.

Sound Made Visible

Cymatics, the science of how sound affects matter, is a technique that makes sound waves visible and helps us understand the effects of sound. Vibrating oscillations of a tone that is sounded beneath a steel plate strewn with quartz sand will cause the sand to form shapes that make the frequency of this tone visible, confirming the effects of vibration on the formation of the physical world which is condensed energy. The shape each tone produces in the sand is constant no matter how many times you repeat this experiment. [81]

Hans Jenny MD, a Swiss scientist, medical doctor, researcher and artist, whose studies led him to examine the effects, structure and dynamics of waves and vibrations in the 1960s, coined the term *cymatics*. His studies greatly enhanced our understanding of the impact of sound. Inspired by 18th-century German physicist and musician Ernst Chladni, who demonstrated the effects of vibration on solid matter, Dr Jenny created more sophisticated techniques and experimented with other materials to make natural, musical and human sounds visible. He found that patterns made by sound bore strong resemblances to patterns found in nature, inspiring the concept that nature itself is being generated by force fields of vibrational energy. This is a concept the scientist Albert Einstein furthered, stating that the field governs matter. [82]

These results resonate with the biblical statements, *In the beginning was the Word,* [83] and *The Word became flesh.* [84] Replace *Word* with *Sound* and we're getting to a closer understanding of the nature of our reality.

One of Jenny's experiments was to expose lycopodium powder to a vocalization of the ancient Sanskrit word *Aum,* thought to be the primordial sound of creation. The vibration of this sound caused the shape of the Sri Yantra to emerge. [85] The Sri Yantra is a 12,000-year-old symbol considered the mother of all Yantras, (geometric shapes that hold the vibrational imprints of sacred energies). Its mathematically precise design of 43 balanced triangles is based on the Golden Ratio. [86]

Herein lies the power of symbols. Each holds the entirety of its meaning and the blueprint for its manifestation. The logo on your cereal box may not carry much power! But the Sri Yantra symbol certainly does. The symbols that carry power are the ones that are reverse-engineered from resonant primordial sounds, the structures underlying the patterns of nature and the frequencies of states of consciousness also called sacred geometry. When we concentrate our attention on these powerful symbols, we are entraining our consciousness to their frequencies.

Another powerful symbol is the Flower of Life. Its complex geometry is thought to be the foundational pattern of all form. The symbol has been found carved in ancient synagogues in Israel, temples in Japan, China and India and in Turkey, Italy and Spain as well as in Phoenician, Assyrian, Indian, Asian, Middle Eastern and medieval art. The great artist and inventor of the Italian renaissance, Leonardo da Vinci, studied and drew the Flower of Life. Its oldest appearance is a carving that was burned into the stone at the Temple of Osiris in Abydos, Egypt, anywhere between 6000 and 12,500 years ago. [87] Symbols that emerge around the world are fascinating. The exact same symbols seem to have been apprehended from the unified field by people who had no physical contact with one another.

The vibration that the Flower of Life symbol embodies is thought to be the underlying formative geometric template of our Universe. Contained within it is the symbol for the Kabbalistic Tree of Life and another informational symbol called the Fruit of Life, which in turn becomes Metatron's Cube containing the five platonic solids, whose geometric patterns emerge in both animate and inanimate life forms as well as in music and language. The geometric template of the Flower of Life forms the basis for everything from the galaxies, to flowers, to the body and proceeds from the Seed of Life and the Egg of Life, which mimics embryonic cell division. Amazing when we consider these templates were perceived prior to the invention of microscopes.

The Memory of Water

Professor Masaru Emoto was a researcher who studied the effects of frequency of thought and emotion on matter. His experiments have helped to clarify the degree to which our thoughts and emotions create frequencies that affect matter. He applied emotions to water in a variety of ways and then photographed the effects on the frozen water molecules. Water molecules of a polluted river or water from a glass with the word *hate* taped to it looked like tiny, misshapen vomits. Water from the same sources when subjected to prayer, or the words *love* and *gratitude,* took the form of exquisite, harmoniously formed snowflakes. Emoto's work visually confirms the effects of the frequencies of emotion on the shape of matter. [88]

We are approximately 70 percent water. We know our thoughts are energetic fields transferring information. So, imagine what it does to us to continually criticize ourselves or our loved ones or others. Imagine the positive effects on our biology and those near us when we are in harmonious positive states and sending loving thoughts.

Emoto's approach is a concrete way to look at the effect of thoughts on manifestation. Let's assume, based on this research, that if we think thoughts about others they will also be affected. Their molecules will shape-shift when the frequency of our thoughts reaches them. In the unified field of energy and consciousness, there is no time, our thoughts travel instantly and find their mark. *Our thoughts have energy.* Energy is a container for informational data and consciousness. It has memory. It is now possible to see with our own eyes that loving emotions are blessings endowed with transformational ability, and hateful thoughts are no different from a curse.

Bernd Kröplin PhD and Regine C Henschel have advanced Emoto's research, creating controlled studies designed to study the effects of ultrasound, Wi-Fi, mobile phones, vibrational therapies (such as homeopathy), music, emotion and cognitive thought on water, further confirming the effects these frequencies have on the water structure within our cells and how they affect whether or not we are of 'sound' body and mind. [89]

In our overstimulated world, it is difficult to achieve the level of deep meditation where peace is found. We are at every turn bombarded by discordant music, discordant thoughts, a continual flow of bad news (good news should get equal coverage), Wi-Fi, social media, a continual flooding of radio waves and thought waves impacting our subtle fields, making it difficult to achieve the clear mind upon which a new story may be written.

Silence is truly golden, so finding time to switch off the noise is important. Walks in nature are an ideal way to recalibrate our resonant fields with those of Earth and are very healing. We can't help but feel awe when faced with nature's beauty, an emotion that quiets down our default-mode networks, the measuring and planning part of our brain, allowing our hearts to open. Recent studies have confirmed that walking in nature improves general well-being, mental and emotional health and physical health and also heightens creativity. [90]

We are musical beings emitting sound frequencies continually. The more evolved, peaceful and harmonious we are, the finer the frequencies we emit become and the further our frequencies travel. Potentially, any of us could positively affect the whole cosmos, as the travel of energy is limitless.

What's Going on in Your Cranium?

The brain is an electrical organ with ten billion interconnected nerve cells (neurons) discharging electrical activity in the form of brainwaves. [91] These brainwaves and their patterns can be observed with an electroencephalograph. If someone is in a coma, they have no brainwave function, whereas someone experiencing a seizure flickers at wildly high frequencies, basically experiencing an electrical brainstorm.

Much as we can communicate by sending the sound waves of our voices to meet an ear that interprets them as information or the radio waves of music can be picked up by a radio device that interprets them and makes them intelligible, our neurons communicate by sending synchronized pulses carrying information that will be interpreted as thoughts, emotions, behavior and physiological functions.

Our brainwaves vary in depth and frequency and like musical notes produce various scales – the lower frequencies reverberate like deep drumbeats and the higher ones like flutes or violins. We are tonal beings in a tonal Universe. Not only are we part of a great symphony with our unique tone in the orchestra of life, there is a symphony inside our brains as these lower and higher waves harmonize and interweave. This music is usually imperceptible but some deep meditators have reported hearing these sounds.

Like other frequency waves, brainwaves are measured in Hertz (Hz), cycles per second. Not surprisingly, slower waves dominate when we are tired, and faster waves when we are alert. Our brainwaves shift with our perceptions. When our brainwaves are out of balance, our emotions will be too, and vice versa. Instabilities, hyper-arousal and under-arousal can lead to anxiety, depression, anger, aggression, insomnia, nightmares, panic attacks, ADHD, obsessive–compulsive disorder, bipolar disorder and a host of other issues including physical symptoms such as hypoglycemia (low blood sugar). [92]

Frequently these problems are dealt with by prescribing medication for discordant brainwaves. They are also often an unconscious impetus for taking recreational drugs. A combination of meditation, breathing exercises, trauma release, listening to harmonious music, and increasing physical exercise can have more lasting, life-enhancing and transformative effects. Qigong, Tai chi, dance, yoga, running, working out, aerobics and other forms of physical expression have been found to be extremely effective in reducing depression and anxiety and can be another form of alchemy on the journey of clearing and transmuting suppressed emotion. Controlled studies have revealed that three bursts of high-intensity exercise per week in 30-minute sessions can significantly reduce major depressive disorder (MDD) and can be as effective as medication. The exercisers showed less relapse than those on medication and were 50 percent less likely to be depressed at their 10-month assessment. [93] The more we move the better. Look at children. They are non-stop motion. We begin to nail ourselves to desks and lose this crucial element of our well-being. If you are on medication, please consult with your health-care practitioner before making changes.

Music is also a powerful tool for entraining brainwaves to optimum function. The music of Handel, Bach and Vivaldi is especially potent for shifting brainwaves. Shamanic drumming can bring people into Theta states where they journey to other states of consciousness. The visionary pineal gland is just above the palate and will powerfully respond to the emotion and vibration of devotional singing and chanting.

Working with the techniques of Creative Alchemy, particularly the first steps of emotional release, I have seen many people, children and adults, find relief from anxiety, depression, anger, aggression, insomnia, nightmares, panic attacks, ADHD, obsessive–compulsive disorder, bipolar disorder and other problems, confirming for me the role that suppressed emotion plays in discordant brainwave function.

A Brief Description of Brainwaves and Their Functions

Infra-Low (<5 Hz). *Infra-Low* brainwaves appear to have a role in brain timing. They are very slow and hard to detect, therefore not much is known about them.

Delta waves (.5 to 4 Hz). *Delta waves* are associated with deep relaxation and deep sleep, and they help to regulate digestion and heartbeat. If we experience enough Delta while sleeping, we wake up refreshed. Too much Delta can cause problems with cognition and cause ADHD. Too little can cause poor sleep and problems with revitalizing the brain. When the brain rests in Delta it is functioning well, we have healthy immune responses, enjoy deep meditations and get restorative sleep. These brainwaves are low frequency and reverberate deeply, like a drumbeat.

Theta waves (4 to 8 Hz). *Theta waves* are dreamy brainwaves associated with a twilight state where we may receive strong imagery or extrasensory perceptions. When we are relaxed or asleep we produce Theta waves. Creative, hypnotic and intuitive states are characteristic of this brainwave which also kicks in when we are in receptive states, such as meditation or learning something new. This particular frequency range is involved in daydreaming and sleep.

Theta waves are connected to emotions and empathy, although too much time in Theta can create depression. It's all about balance. This is a powerful state of consciousness.

Alpha waves (8 to 12 Hz). *Alpha waves* create a calm, peaceful, integrated, coordinated, resting and relaxed state for the brain and body. It's a state of low arousal and a good one for learning, analysing and performing elaborate tasks. It's the frequency between Theta and Beta, therefore it bridges subconscious and conscious activities. If we have too much Alpha, we are too relaxed. Too little and we are anxious. In balance, we are very present.

Beta waves (12 to 38 Hz). *Beta waves* kick in when our minds are engaged in cognitive tasks, including activities like musing, problem-solving, making decisions and complex thought and help make us highly alert. If we have too much Beta, we can't calm down – we'll have insomnia and monkey mind, tension and anxiety. Engaging in conversation is a Beta activity. This brainwave function is geared to the outside world whereas Delta, Theta, and Alpha are more internal. Beta uses a lot of energy and is therefore best for shorter bursts of concentrated activity.

Gamma waves (38 to 100 Hz and beyond). *Gamma waves* are also involved in higher processing and cognitive function. These are involved in higher processing tasks, such as information processing and learning and general cognitive functioning. People with low Gamma brainwave function can have trouble learning. Gamma brainwaves are associated with the transcendent states of unity, altruism and unconditional love. They harmonize the body, mind and spirit and relate to expanded awareness. These are our fastest brainwaves and sweep across the brain, integrating perceptions while simultaneously processing information from different brain areas. We must be very still to access these subtle brainwaves. They are characteristic of advanced meditators and martial arts masters.

A healthy brain moves through many frequency states easily with great flexibility, like a virtuoso musician playing a highly calibrated instrument. Meditation is excellent for encouraging steady optimum brainwaves.

The Sound Waves of Our Food

When I was living in London I had the good fortune to come across a healer and energy worker with great expertise and knowledge. He was quite elderly and had gained much wisdom in his 50 or 60 years of practice. He had very advanced Russian technology that could balance the energy fields of the body magnetically, and technology from the flight panel of the Concorde airliner, a supersonic aeroplane that could travel at twice the speed of sound that operated from 1976 to 2003. He had a very interesting tool that could measure the frequency of anything. I'm sure in the world of science this is not unusual, but for me it was quite fascinating.

I had been eating top-quality organic food and felt awful. I have a high level of sensitivity and despite the quality of the food it felt dead to me. I brought various examples of the food for him to measure. It was all close to zero Hz! We pondered how this could be. The food had been bought from stores that were part of large organizations. It's possible it could have been radiated for longevity, therefore killing it. We also wondered if ethos could affect vibration and therefore Hz.

This is an ongoing issue for all of us. Food as close to ideal growing and harvesting conditions is important. I now get my food from small outlets providing organic food from local small farms. If I am with friends or family, I feel food lovingly prepared outweighs provenance. Praying and decreeing over our food significantly expands the frequency and raises its life force.

Mantra

You can create a shield of brainwave-altering sound power by chanting words that gradually change your frequency and offer ballast against the turbulence of the competing frequencies around you. Mantra is the meditative, contemplative repetition of the very high-frequency phrases. When we receive a mantra directly from an enlightened master it can instantly alter our consciousness.

If we do not have an enlightened master to hand, utilizing mantra sincerely and continually can still gradually change our entire being, the neurological makeup of our brain, the energy fields of our mental, emotional and physical body and align us with the supreme consciousness of our Presence. Just as cymatics can rearrange the quartz sand crystals on a metal plate and emotions rearrange Professor Emoto's water molecules, mantra can rearrange your frequency pattern. Mantra is a powerful way to create new positive neural pathways.

The greatest Guru is your inner self. Truly, he is the supreme teacher.
He alone can take you to your goal and he alone meets you at the end of the road.
Confide in him and you need no other Guru.
– Sri Nisargadatta Maharaj

Mantra is not confined to any one culture. Repetitive prayer is also a form of it and so is decree. We can decree ourselves to be one with inherent supreme qualities such as love, wisdom, power, abundance and joy. Choosing a phrase that has been used for millennia and has enormous potency and *momentum* will accelerate the transformation of your energy field and your inner alignment with your highest potential.

Two of the most powerful phrases to repeat are *Om Namah Śhivāya* and *I Am That I Am.* They both are salutations to the Source of all while acknowledging that Source is also in you. This can be done internally, or out loud, as long as it is continual. The mystics and sages of old, with their understanding of the unified field, the nature of reality and the undivided Self, comprehended that the repetition of high-frequency mantras such as these could lift and transmute lower frequencies by entrainment (the synchronization of multiple waveforms, in this case from lower frequency vibration to higher). Singing your chosen mantra is also extremely effective.

Sound is energy made audible by vibration. Everything is made of energy; everything vibrates and everything has a frequency. Those who practice mantra receive the benefit of the power of mantra, entraining their own consciousness to the frequency of the mantra. With devoted practice, mantra can dissolve that which separates you from your true essence. Your true essence is the part of you that is one with the 'all-pervading Name,' the ground of being that presents itself in myriad forms, including our own, the highest consciousness attainable, the Presence of all life. [94]

Perhaps most importantly, mantra and comparable practices enable us to replace the negative tapes we play over and over in our heads, often on an unconscious level. In my personal experience, the use of mantra along with the inner recitation of certain prayers has been a huge assistance in completely healing from chronic anxiety. In fact, with sincere practice you will eventually find yourself waking in a state of bliss you can consciously increase until you melt!

Vocal Toning

Vocal Toning is also called Nada Yoga or Sound Healing. Toning harmonizes your vibration. Special ancient seed sounds and vowel sounds are used to clear the 72,000 energy pathways of our being which may be blocked. When energy pathways are blocked, they can interfere with the flow of life force causing anxiety, stress and eventually physical illness. Sound baths with singing bowls and gongs also come under this category.

Aum (Om) is a Sanskrit root word or seed sound symbolizing the aspect of Godhead that creates and sustains all things, the Cosmic Vibration synonymous with '*In the beginning was the Word.*' Taking a deep breath and exhaling with a hum is also very calming and restorative. Direct it through each one of your energy centers for a beautiful sound bath. Really accentuate the three syllables and feel where they resonate in your body.

The Science of Decree

A decree is a statement or command that utilizes the words I AM. The phrase I AM, as it was first given to humanity, is called the *Word of Power* or *God in action* and brings to bear the pure animating force of the universal I AM Presence. When coupled with emotion, will, vision, action and intention, these words activate our blueprints for manifestation. When we learn that the power to manifest is within us, we will no longer feel the need to take from others or feel short-changed. Our supply is within. Understanding the science of decree is crucial to learning how to consciously create your life and manifest your highest destiny and desires.

Any statement that follows the words *I AM* sets the universal Presence in motion. How often have we said *I'm broke, I'm tired, I'm fat, I'm getting older,* or *I'm unsure I'll ever succeed at manifesting my dream*? Each time we do, the Presence receives a docket and goes into motion creating what we wish to manifest.

We can reverse this by using decrees as we use affirmations, only for positive transformation. Decrees, however, hold infinitely more power. A decree qualifies the substance of our being, literally changing our energetic imprint. One of my favorites is *I AM Cosmic Victory*. Others you can employ are *I AM Infinite Abundance, I AM Immortal Love and Victorious Peace, I AM Perfect Health, Wealth and Joy.* After a while you will begin to feel the power and your energy field shifting. To hand the power to your true essence, your undivided Self, that part of you that holds your highest destiny, your I AM Presence, just state *I AM the only Presence acting here.* We will delve more deeply into the science of decree in Chapter Five.

The Power of Prayer

Authentic prayer is as unique as kissing and also a form of communion. In the King's Chamber at the Pyramid of Giza, a disparate group of spiritual travelers, tourists, children and guards had fallen into spontaneous toning the moment the configuration of stars signalling the opening of a portal in time–space had formed in the sky, all holding hands, strangers unified by unearthly, exquisite echoing sound.

Later that evening, drawing a group into a circle deep in the desert under a veil of stars, each participant shared a prayer. The most beautiful was a prayer sung by an Israeli girl who had never been asked to speak in public before, let alone pray, and was moved to tears, as we were. They were all beautiful and heartfelt as we went one by one around the circle. Leading by handing over and supporting is powerful and rewarding, a new paradigm practice that is influencing education and business.

A holy man with shining eyes arrived from the great expanse of the desert and confirmed the relevance of the shift we were celebrating, pointing to the formation of stars in the sky that formed a triangle facing up over a triangle facing down, like the Star of David, signifying 'as above, so below.' He took the three of us who remained still sitting silently under the vast canopy of shimmering stars to an ancient tiny pyramid where other holy men had prayed for millennia, led by a man with a camel. The energy was palpable and a quiet awe descended as we crawled into the small space. We asked where the man with the camel had gone and were told he remained outside to pray for the complete opening of our hearts. Somehow that was the most touching and humbling moment of the day.

Prayer is the key of the morning and the bolt of the evening.
– Mahatma Gandhi

Prayer is charged energy. It's light, sound and electromagnetism translated into electrical and chemical signals within our bodies, which travel on beams of light to those to whom we send prayer. An ancient prayer such as the Lord's Prayer, which emerged from the heart of Judaism into Christianity, or any other spiritual decree from any culture in our world that has been said with passion and sincerity down through the ages, can carry great power because it has been repeated with love millions and millions of times, building profound gravity and momentum. For me, the Lord's Prayer is the story of aligning ourselves with our Presence to co-create heaven on Earth, our real job description, starting with our own personal heaven and building outward.

The practice of prayer expands our communion with the Source energy of the Universe as well as the indwelling Presence in us and in all humanity, for they are one. Prayer unites head and heart, performing as a special high-precision tool which produces specific outcomes by entraining our frequency and reprogramming our consciousness. I think of God as the forces of life and love, for *the will of love is life and the law of life is love.* Devotion in motion. God is the universal Presence permeating and enfolding all, the infinite, expansive creativity from which flows the sacred, architectural geometries of universes and galaxies, lovers and daisies and all the rivers of evolution that continually transform life. When I pray, or meditate, the experience is sensate, which is how I know I have connected. I feel it in my body first.

Prayer is a communion exercise done to connect to universal qualities of being, such as love, wisdom, power and abundance that we can bring to the forefront of our lives with practice. A habit like this helps us build momentum towards our final merging with the Presence.

The meaning of the Lord's Prayer became revealed to me through repetition. It is a mantra and decree to solidify the blueprint for a Golden Age on Earth. Like the messages in the water experiments of Masaru Emoto, our words in combination with the emotions that lie behind them tangibly affect our world. Like Hans Jenny's experiments, intoning a series of sounds has the power to reshape matter. This is the science behind affirmation, prayer and decree. Prayer practice is very effective for decreasing and finally eliminating anxiety. The degree of sincerity and heightened emotion we are feeling as we say a prayer are significant factors. Prayer doesn't work if we parrot the words, though. We need to really feel into them and then the inner mystery is slowly revealed. There are stages to illumination and prayer oils the wheels of progress considerably. It keeps us humble while true worthiness and confidence grows. I use the original, ancient Lord's Prayer but this is the meaning for me:

Beloved Mighty Magical Presence,
The Wisdom, Love and Power of the Universe,
Of humanity and of all Life,
I align myself to your Infinite Eternal Consciousness completely
That I may create a heaven of this Earth,
A temple of this body.
With deep gratitude, I receive from you
My bread and my breath.
You behold me in my perfection
And inspire me to do the same for others.
Your Law of Forgiveness heals all separation.
Guide me in compassionate grace.
Protect me always in the sacred fire of your Radiant Light.

For you are the worlds within us and around us,
The wisdom that orchestrates the music of the spheres,
The power that manifests the dream of our destiny,
The love that holds consciousness is form,
And the glory of Life itself.
And we are One.
So be it.

When I connect to my Presence by saying the Lord's Prayer or through another intentional practice, I am sometimes wrapped in bliss beyond description and inspiration and insights often come. To move into a deepening sense of connection to the unified field that is permeating, empowering and holding us all, sensing the Presence popping in for a star-filled cup and hugs of bliss-filled energy is thrilling. This is a relationship available to you, too.

Summing Up the Power of Sound and Energy

We are energy beings emitting frequency and vibration, which form patterns that affect the manifestation of the world we live in. This world is a sea of energy also emitting frequency and vibration that form patterns which affect us personally. Energy is information and energy fields are continually communicating information.

When we are walking through our lives hypnotized and unconscious, we are in a state of *cognitive dissonance* and unable to see the bigger picture. We are stuck in our mental prison of limitation. Anything outside of our frame of reference is considered weird or even dangerous.

As we heal and wake up, our energy fields quicken causing us to become aware of the deeper patterns around us and within us and aware of the interconnectedness of all things. It becomes difficult to engage in abusive behavior towards ourselves and others and finally impossible. We become freer and freer of limitation. We access higher aspects of our being and see the cause and effect not only of our actions but also of our thoughts and emotions. We begin to understand the importance of harmony on all levels but also understand it can't be achieved by putting a band-aid on a volcano. We have work to do to clear the hidden aspects of our personality. We will find, in time, there are jewels in the shadows and strengths in our perceived weaknesses. As we express what is buried we are less triggered and have more compassion. We find we are more inspired and solutions for problems come easily. Our song becomes resonant and clear and our pitch on-key. We become part of a great symphony – in the flow, in tune, sound of mind and body.

Conscious creative collaboration is a hallmark of our new frontier. Elevating our consciousness is the most important thing we can do right now as well as right action. When a good proportion of Earth's population is willing and able to apply these concepts we will be able to co-create a Living Vision of a healed and whole Earth and manifest it together. The scientist Rupert Sheldrake says the figure we need to create an evolution of thought is just ten percent. We can do it!

- Nature resonates at a very high frequency (Hz). Just as a tuning fork entrains and tunes a piano, walks in nature can restore our own frequency to harmony. Nature can also 'wash off' the electromagnetic frequencies less harmonious for our well-being such as too much exposure to Wi-Fi.

- Listening to harmonious music is excellent. By the way, please rock out with your favorite music. Wild expression is wonderful. If you are feeling stressed or anxious it's worth looking at what you are listening to and perhaps experimenting with some different sounds.

- Using techniques such as Mantra, positive affirmations, prayer or decree as a running tape is a very good way to create new neural pathways of harmony and beneficence. Ideally you feel into the meaning of the mantra, which increases its power and engages more of you.

- Toning and chanting are wonderful ways to entrain your energy field and physical, emotional and mental bodies to a higher frequency. Higher frequency states are expanded, inspired, relaxed and harmonious. If you are trying to solve a problem, taking some time out to shift your brainwave frequency is a good start. Just take a deep breath and hum softly on the exhale. Place your hand on your heart and think of what you love. Expand into gratitude and then tackle the problem!

- Sing songs that you love, sing in the shower or when tidying the house.

- Keep a journal and note what your environment is when you are feeling optimum and when you are feeling stressed. How much is attributable to your outer environment and how much to your inner environment?

- Walk barefoot on the Earth when feeling ungrounded or confused.

4

OUR QUANTUM MANIFESTOR – THE POWER OF IMAGINATION

The task of the 21st century is devotion to the visionary imagination.
– Jean-Yves Leloup, theologian and philosopher

The mental body is our *quantum manifestor*. The images of our mind have the power to create. These visual ideations form strong geometric patterns in the unified field especially when fueled by emotion, influencing our reality. Our ability to visualize is a crucial component of conscious manifestation. If you are currently unable to, there are special techniques in Part Two do develop this mental muscle. When our mind is in alignment with our Higher Mind, empowered by visual imagery and strong healed emotion, we become powerful architects and builders. Our imagination becomes much more than a flight of fancy; it is transformed into *Living Vision.*

To elevate imagination to Living Vision capable of consistent manifestation, we must also connect to the electricity of our Presence and its infinite force field. The Presence is anchored in our hearts where the flame that does not burn but transmutes all it touches, the eternal Threefold Flame of love, wisdom and power, is anchored. This is the love that magnetizes and coheres, the power that can imbue life and the wisdom that illumines. Manifesting with the techniques of Creative Alchemy is an alchemical science of the heart.

The great works of our Earth have emerged from the Living Vision of their creators because those individuals had found their alignment. These are always works of enormous devotion, illumination and power.

You Can Become a Magician

When our emotions are healed we are in a position to activate our latent superpowers. We can become masters of our reality instead of victims of it. We can consciously create the images that inevitably out-picture into physical form instead of unconsciously creating discordant realities which cause us to believe life is 'happening to us' and that we are victims of it. We are the author of our lives. We can rewrite the story at any time.

The power to influence events using what has been considered mysterious or supernatural forces was once considered magic. Many of these forces are now being demystified by modern science. Today, as we come to understand more about how the mind and emotions work and how our energy affects our reality, what was once considered magic is seen to be within natural law.

Early practitioners of systems designed to affect outcomes and influence the nature of reality include wise men and women who intuited remedies from plants, many of which form the basis of medicines we use today; the priesthood of the Egyptian pharaohs who used empowered imagery and heightened emotion to influence manifestation; practitioners of nature magic who understood how the ritually summoning the forces of nature could empower manifestation and mystics of old who understood how the patterns of energy and light affect reality, knowledge that is just now being scrutinized in theoretical physics. Occultists and adepts understood they could impose an energetic blueprint, a symbol or image on the quantum field and influence the patterns that create physical form. It is possible that this knowledge and understanding was behind the building of the pyramids of Egypt and the famed Temple of Solomon in Jerusalem. Individuals who had developed incredibly strong powers of visualization could employ a force called *levity*, the opposite of gravity, with technology of mind to move heavy stones. Some summoned the force needed to enliven the blueprint with 'sex magic,' a way of marshalling the forces of life to animate an image. Masters of the mental energy field knew how to use their own energy body to conjure an image of an 'other' that could be sent in the form of subtle energy to do their bidding.

We are relearning the power of combining images with focused will, intention and heightened emotion and then supplying the electrical animating voltage and illumination by connecting to Source energy, the living consciousness that permeates and enfolds all things. It can be brought up from the Earth or drawn down from the Sun or the center of the galaxy. Ideally it is finally drawn from deep within yourself, from the place deep within your heart where your Presence connects with your mental, emotional, physical and energy bodies. This is exactly what you will learn to do with Creative Alchemy.

The new understanding of the responsiveness of subatomic matter and the implications of how our consciousness interfaces with the fields of energy permeating and enfolding us can give us a sense of freedom. We are not imprisoned by our world's reality; we have co-created it. We can co-create it anew. When we begin to understand how truly creative we are and how we are creating realities in every moment, we must also develop a greater sense of responsibility. The great power of our imaginations and emotions can be combined for either our good or detriment! We are unconscious magicians on our way to becoming conscious magicians through the process of Creative Alchemy, the transformation of lead to gold, victim to master. We live in a multidimensional sea of potentiality. The one we focus on will become a physical reality. We are responsible for everything we have created, even the thought forms that have not made it into form because we've given up and what we create in our dreams while sleeping, a sobering thought.

We can now return to the ancient knowledge of magic that has been unearthed around the world from various traditions and sources and, understanding it through the lens of modern science, seeing its validity. New science has become our friend as we reclaim our birth right as powerful magicians able to create our own destiny and co-create a better world. It becomes clear that being a magician is actually our natural state.

The Flexibility of Our Reality

Just how flexible is this reality we inhabit? The Toltec shamans tell of whole worlds they have the ability to create and inhabit. They achieved this through rigorous techniques of harnessing energy and empowered 'dreaming.' Conscious dreaming is a self-created movie written, directed and produced by us that can become reality. All creativity and manifestation is intimately connected to our stores of energy. It is our currency. So, what are these realities we inhabit and create?

Many innovators and scientists including Elon Musk, Neil deGrasse Tyson, Gregg Braden and Nick Bostrom believe it is possible we could be living in a simulated reality – a sort of giant Virtual Reality (VR) game, perhaps one created by our ancestors or even our future selves. According to the *New Yorker* profile of venture capitalist Sam Altman, two tech billionaires have engaged scientists to work on breaking us out of the simulation![95]

Brian Whitworth offers the possibility that the quantum world is real and generates the physical world as a Virtual Reality. This suggests our physical world is a dream emanating from an absolute reality. This is certainly what saints and mystics have said. There is an eternal reality and a temporal one. The temporal one is a dream or cosmic play – a Virtual Reality.

In this virtual world, you could take on any historical personage you wished as your game avatar, checking out the 'book of data' from the universal simulation library that stores all the particulars.[96] Perhaps this is the reason why so many people believe they were Mary Magdalene, or Jesus or Joan of Arc. They've all checked out the same book and are assuming the same avatar.

I believe if we are in a VR setting, it is just as likely that *we* are the ones generating it in every moment consciously and unconsciously. Our multidimensional Presence could be creating an avatar of our four earthly bodies and is 'playing us' in an incredibly sophisticated game to learn the rules of manifestation in a slower, denser dimension. When we have restored our gifts of *technology of mind,* we ourselves can break out of the simulation and create another. Perhaps it is all a dream within a dream, but at least we know we can create our own dream and not settle for the one imposed on us. Creative Alchemy is a tool to dissolve the illusion that has us hypnotized in a dream of limitation and unworthiness and replace it with a dream of limitlessness, freedom and sovereignty. Together we are becoming the dreamers of a new world of infinite potential, abundance and well-being. It's my dream to be all I can be. What's yours?

Imagine this. You are a great spirit of fiery energy, pure life force, issued forth from the Infinite Source of All That Is. You are its 'mini-me,' a fractal aspect of the universal Presence imbued with all its qualities but also your own unique essence. You are a creator being made in the image of your Source *by way of your own ability to imagine and create.* Your ability to project your vision and form a reality by the applied use of your elevated imagination, will, passion, action and intention is your superpower.

Let's assume you *are* in a Virtual Reality. In this case you don't push a button to change your scenario, you employ your imagination. The imagination when empowered as Living Vision can create worlds.

You are a being of stardust, bioelectrical energy, and love, who is eternal and infinite. In your avatar, you are attending Earth university and part of the game is amnesia. As you score points through insights, realizations and successful manifestations you gradually start to remember who you really are. The veil thins. You are learning how to negotiate limitation and then free yourself from it. As you start to excel at the curriculum, it becomes really exciting and multi-layered. Once you have used the beginning techniques of Creative Alchemy to heal yourself emotionally, the fun can begin. When you find your true alignment with the deep eternal part of yourself, what is most meaningful for your journey will unfold. And it will be a joy!

You are a creator being hanging out in one of the infinite multitudes of dimensions, whichever one the frequency of your energy field matches. *You* decided that you wanted to experience this beautiful and slightly crazed emerald and blue globe everyone's been talking about. Your frequency found a match in the perfect parents, your doorway to physical form and the match for the life you needed to craft, the perfect life to learn what you needed to learn and experience on your path to mastery. Your Presence projected the Living Vision of an energy template, like an architectural blueprint, with infinite love, power and wisdom, particular capabilities, challenges and potentials.

When it passed into the Earth realm, this energetic blueprint began to coalesce the whorls of energy that became your seven main energy centers and they began their work as transducers of universal life energy into bioavailable energy. Then your father's spermatozoa hit your mother's ova and voila! The sacred geometry of cellular division began its miracle. This cellular division mirrors the Golden Ratio and Fibonacci sequence, creating patterns beginning with the Vesica Piscis. It continues to divide and reform, transforming into all the geometric stages preceding and including the Flower of Life, on its way to forming your quantum navigator, your physical vehicle for the Earth journey; your quantum generator, your emotional body; your quantum conductor, your dream body or subtle energy body; and your quantum manifestor, your mental body.

Your challenge is to transform the four earthly bodies from lead to gold, elevate your mental body to your Higher Mind and your imagination to Living Vision and enter the portal of your Presence anchored at your heart – the *only* consciousness that can act in your life. With that you graduate to your master's degree, aligning with your true purpose, manifesting your unique expression and co-creating a conscious evolution of our world.

For this remarkable feat you are allotted a certain amount of life substance from your eternal store. You have free will to clothe this substance in harmony or discord but you will have to eventually return it cleansed. The greater the harmony you can achieve the more abundance you can consciously sustain, store and release. Harmony and gratitude are keys to your account in the universal bank.

Whatever challenges you have had to experience in this life are part of a role you have assumed and that you will one day shed. This greater eternal part of you is ecstatically singing your unique tone in the realms of perfection while a finger of yourself is having a massive headache in our particular space–time corner of the universe. The more we can focus on the greater part the more we can merge with it and the game changes.

The final step, your doctoral degree is the integration of the Presence, the unique you in your most glorious manifestation. This is what Creative Alchemy assists in birthing.

For millennia the spiritual trajectory has been to get off the wheel of life. However, as more of us awaken, it's becoming clear we can create a heaven of this Earth, a paradise, by shifting our perspectives and perceptions, opening our hearts and doing the work. The divine is everywhere. We don't need to leave Earth to find it.

Earth's crammed with heaven
and every common bush afire with God. [97]
– Elizabeth Barrett Browning

You, yes you, reading right now, are majestic beyond reckoning. You have hidden your light to conform, to be part of the status quo, but you're going to release it now and unleash your power and really shine. You don't need to become anything; you just need to peel away the layers hiding who you really are. That's what Creative Alchemy helps us to do. But we're going to take it *sloooowly*.

As we continue to integrate and align our minds and hearts, we become ready to activate our imagination. Those castles you've been building in the sky can land. We can become true conscious creators, able to bring back treasures from imaginal realms to build a paradise of our lives and the lives of others.

What Are the Imaginal Realms?

The imaginal realms are aspects of reality not held in place by the 3D structure of time and space called by scientists the space–time continuum that defines our current consensus reality. They are no less real. Quantum theorists are identifying many dimensions and this is what the imaginal realm could be called, a different dimension from our own, separated only by frequency. In this world, the imagination is an agent of cognition and the architect for *manifestation*, not of fantasy.

English theologian and 18th-century Romantic poet Samuel Taylor Coleridge stated that imagination was the condition for cognitive participation in a sacramental universe – a means of establishing a cognitive and visionary relationship with an intermediate world, a world that is more fluid than the one we usually perceive, a timeless world. He states, 'The primary Imagination I hold to be the living Power and prime agent of all human Perception, and as a repetition in the finite mind of the eternal act of creation in the infinite I Am.'[98] This primary imagination is what I call Living Vision. He goes on to differentiate this living power from mere fantasy: 'The Fancy is indeed no other than a mode of Memory emancipated from the order of time and space ...'[99] a brilliant summation.

Twentieth-century physicist Albert Einstein spoke of realms outside of time. He believed time was 'numerical change in space' meaning time is relative to ordered events and not an absolute.[100] The theoretical physicist Carlo Rovelli reminds us that when we used sundials we were closer to accurate measurements of the sun's journey and our arbitrary mechanical clock divisions are not in sync with the natural ebb and flow of time.[101] Our imagination is a kind of interdimensional space ship that can bring us backwards and forwards in time and space to create and even change events. How far can this go? Russian scientist Grigori Grabovoi is doing some remarkable work with his students, accessing universal mathematical codes to move time and have a direct impact on the regeneration of the physical body. [102] The relevance of these perspectives to Creative Alchemy is that we will use timelessness to travel in consciousness to the past and to the future, and deep into the eternal now where our Presence dwells.

Higher dimensions are actually bands of frequency, octaves that can be negotiated via individual consciousness according to our personal levels of frequency. Modern science has its own theories on multi-dimensions. Theoretical physicist Michio Kaku uses the term *hyperspace* to describe the multiple dimensions he believes form our universe, a theory he believes also explains black holes and the potential of time travel. He says the laws of physics begin to make sense when we allow for hyperspace, stating matter is a vibratory field rippling through the fabric of space and time. [103] He is the co-founder of string field theory, which he believes is the 'theory of everything.'

> *Everything we see around us is nothing but the vibration of tiny strings.*
> *Each subatomic particle is a note vibrating on a tiny string. What is physics?*
> *Physics is the harmonies we can write on vibrating strings. What is chemistry?*
> *Chemistry is the melodies we can play on vibrating strings. What is the universe?*
> *The universe is a symphony of strings. What is the mind of God? The mind of God*
> *is cosmic music resonating through eleven dimensions of hyperspace.* [104]
> – Michio Kaku, theoretical physicist

I love this quote. Esoteric wisdom schools have long identified 12 dimensions in our universe, each with many striations. Science is almost there!

In Part Two, the practical application of Creative Alchemy, we are going to play with time and space in the imaginal realm. This imaginal realm is a dimension we can visit using our imaginations, a place where we can go to retrieve and build our dreams and bring them back through the rabbit hole to manifest in ordinary reality. We are going to create quantum events using our consciousness and focused attention, collapsing reality into *Consciousness Observer Moments* wherein our empowered vision can create a new reality through attention and intention. The imaginal realm could be called the precursor to this world we live in, a place where the fabric of existence is far more malleable than it is here. This is where you will go to activate your Living Vision – seed it and 'water' it, tend and nurture it, listening and looking for signs and synchronicities.

During this time of great awakening, the fruits of the imaginal realms can be found enriching life in unlikely places. Psychoneuroimmunology, a new field based on the new understanding of the body–mind matrix, uses the imaginal realm to influence physical healing. The immune system responds to guided imagery and visualization. The body does not differentiate between the reality we believe we inhabit and the reality of the imagination. It experiences both as actualities. This is not new to shamans and mystics who have employed similar techniques to Creative Alchemy through art, music, visualization, dance, psychodrama and other methodologies, to heal mind, body and soul for thousands of years. Eventually, imaginal technologies will become commonplace in the mainstream healing of illness. Activating new realities based on a true understanding of the power of the imagination will alter our perception of our currently latent powers.

Carl Jung brought the wisdom of the imaginal realms into his practice of depth psychology through activities such as active imagining, which is a meditation technique similar to shamanic journeying. Through free-flowing creativity such as automatic writing, dance and painting, symbols of the unconscious mind rise into conscious awareness where they may be translated and interpreted. What are the unconscious, conscious and higher conscious minds if not realms or dimensions? They are vibratory fields rippling through the fabric of space and time and beyond. Everything is consciousness.

Jung came to describe the realm in which he met archetypes and guides as the realm of the *collective unconscious.* [105] Born three decades later, the psychologist Abraham Maslow called the same realm *transpersonal,* relating to experiences extending beyond the personal identity into a wider aspect of psyche and cosmos relating to transcendental experience beyond the self or ego. [106] Our ability to dream, empower our dream, and consciously manifest is our greatest gift except for our ability to love.

> *Dance first, think later. That's the natural order.*
> – Samuel Beckett

The Key to the Treasure Chest

French philosopher and theologian Henri Corbin created the theory of the *mundus imaginalis* (world of images), the kingdom of subtle bodies and archetypal intelligences where human and divine imagination meet, mediating between the world of the Spirit and the body. [107] Corbin, who coined the term *imaginal,* stated that we cannot enter the imaginal realms by housebreaking. I think this is what is meant by the phrase 'You can't storm the gates of heaven,' something I have learned the hard way. The key to the door is frequency, in particular the Theta brainwave state. Theta is a state of relaxed but alert dreaming. Achieving this can be assisted by meditation, breathing, authentic creativity and intention. When you have achieved the state enough times, relaxation, intention and attention are generally all that will be necessary to re-enter.

There is no map I can give you of your imaginal realm, as it is unique terrain. Our paths may cross there but each of us occupy our own. If entered humbly and with an open heart, you can meet master teachers and guides and build the template of your destiny. It is a place to manifest your highest dream of life in the here and now.

> *I am certain of nothing*
> *but the holiness of the heart's affections*
> *and the truth of the imagination.* [108]
> – John Keats

The Role of the Imagination in Sympathy and Empathy

Imagination plays a surprisingly profound role in our social well-being.

Sympathy is feeling sorrow or caring for the plight of another. Empathy is taking it one step further, actually being able to stand in the shoes of the other person. Both of these states are acts of the imagination. It is imagination that allows us to understand how another is feeling and this understanding leads us to charity, tolerance, benevolence and patience. Both are crucial for inspiring the desire to help others and commit altruistic acts that do not benefit us. Both lead us to feelings of compassion, one of the highest states humanity can achieve.

Lack of sympathy and empathy has the opposite effect. One of the main traits of psychopaths is lack of empathy. [109] It makes perfect sense. Without the ability to imagine how someone else feels we lose our humanity. The 'other' becomes objectified. This failure of the imagination and consequences of the lack of it can be seen everywhere.

Eighteenth-century philosopher David Hume postulated the idea that sympathy is behind altruism and the inspiration behind ethics. We are motivated to help others because we can understand how they feel. [110] His contemporary, philosopher Adam Smith, saw how the imagination allows us to truly put ourselves in the shoes of another, so, for example, if we condole with the loss of a friend's son, we enter *their* grief and don't consider how we would feel in those circumstances. We do so as if we change person and characters. Our grief is entirely on their account and entirely unselfish. [111] Twentieth-century psychoanalyst Heinz Kohut saw empathy as the primary method of data gathering in the discipline of psychoanalysis, and also the oxygen in which a healthy child will flourish and grow into a healthy adult with good relationships in love and work. He believed empathy humanized us and without it our experience of one another would be incomplete. It's a form of receptivity. [112]

Are the problems in our world a failure of the imagination? I believe so. There is very little understanding of the potency and crucial need for imagination in our world. The arts, which help develop imagination, are considered secondary to academics instead of an equal partner, as they should be. As well as fostering innovation, invention and the ability to problem solve, imagination helps us become empathetic and sympathetic. The correct application of the imagination can transform stuck and lost individuals into solution-oriented problem-solvers who are inspired to align their purpose with the greater good.

As Artificial Intelligence becomes a huge force in our world, cultivating the ability to innovate and think creatively will become increasingly important.

The Imagination of the World

I believe that semantics and misunderstanding of words and concepts are behind many of the divisions in our world. To imagine is to make images. It doesn't mean to live in a fantasy world. Everything new ever created was an act of the imagination.

The Hindu culture describes the Supreme Cosmic Spirit, Brahman, as the pervasive eternal truth that does not change but underlies all – the Truth of truth. [113] Similar to the esoteric science of the I AM Presence, there is no difference between Brahman and Atman, the individual soul. [114] The ancient Greeks conceived of a world soul, which translates in Latin to *anima mundi*. Plato stated that the world is a living being endowed with a soul and intelligence, a single visible living entity containing all other living entities, which by their nature are all related. [115] 2500 years later scientists are still catching up but the more forward-thinking of them have come around to the idea that Earth is at the very least a self-regulating organism.

Most indigenous cultures experience Earth as a living being, the Mother. It is only one small step further to go from perceiving the Earth as a living being endowed with imagination and that imagination having had a role in the creation of humanity, animals, minerals and plants. Does the collective imagination of humanity interface with the imagination of Nature? Many cultures believe in and communicate with Elemental beings, the subtle energy bodies of the elements, plants and minerals in charge of the templates that precede their form just as our subtle energy body does for us.

Walk in the forest or by the sea and allow yourself to expand your consciousness to connect with the sentience of the natural world and experience what that is like. There are many stories of Nature gifting songs and other inspiration. I have never been musical but spending a month by the bubbling River Cree in Scotland I wrote four songs. I feel they were co-creations with the river. I later discovered that many cultures believe rivers are endowers of musical gifts. There is magic in the pure clear frequencies of Nature. Its purity and beauty are uplifting and bring inspiration which expands the imagination.

Biologist and author Rupert Sheldrake speaks of the cosmic imagination and the imagination of galaxies, planets, ecosystems, societies, organs and tissues. [116] Isn't it arrogant to think we are the only ones imbued with imagination? Animals certainly possess imagination and are able to innovate and problem-solve. Could it be that imagination is inherent in the consciousness that permeates all? Did the sun slowly dream the solar system from a molecular cloud of hydrogen? If the universe as we know it emitted from a Big Bang, what imaginal force created the concise vibrational geometric shapes of mathematical music that led to me in my chair and you in yours? How long will it be until the new physics of today merges fully with the metaphysics of the ancients who saw energy as consciousness itself? Joining the noted physicists who made this leap in the 20th century – Max Planck, David Bohm, Niels Bohr and others – are 21st-century scientists such as Christof Koch, Sir Roger Penrose and Giulio Tononi, who are examining the possibility that energy actually *emerges* from consciousness. [117]

It is conceivable that the larger consciousness of the whole interfacing with the consciousness of our neuronal tissue creates the human mind by introducing the chemical interpreting mechanisms of physiological perception. Memory, that which creates identity, is an act of the imagination. We are a catalogue of memories. How we reimagine our past and what we have been imprinted with is affected by our perceptions. A group of people can remember (reimagine) the same past completely differently. Could our minds be the personalization of consciousness through the agency of memory and cognition?

What does this have to do with Creative Alchemy? Your imagination is one of your greatest gifts. It is the precursor to conscious co-creation and a crucial tool in creating the destiny you were born to live. Honor it in yourself, in others and in the fabric of consciousness itself.

Protecting the Imagination

We live in a sea of thought forms. Many are visible, such as the relentless pounding of the news on TV which is always biased to the sensational and the negative. Where's all the good news? There's plenty out there. Then there are endlessly noisy marketing and entertainment images bombarding us. Thought forms and radio waves flood our mental atmosphere.

The ancients removed themselves from daily life when they wanted to commune with the Presence, knowing that what the mind and body perceive affects what you are attempting to manifest. It is important to find some solitude to develop the deep listening you need to hear your inner guidance and time to recalibrate from the non-stop flood that is modern life.

As mentioned, we have mirror neurons in our brains that respond to actions we observe in others as if they are happening to us. [118] So, while you're watching a horror film your body is releasing crazy amounts of cortisol and adrenaline. This may give you a momentary thrill, but it is training your receptor molecules in a way that will adversely affect the frequency of your energy field. I personally like intelligent psychological horror. I feel it is an excellent vehicle for expressing aspects of the human psyche, but I never watch slasher horror because my mirror neurons have a field day, and the thought forms can continue for some time, triggering the cascade of non-beneficial chemicals along with unrest. For me it comes down to a choice. Which is more important, manifesting my highest destiny or cheap thrills? However, I am an empath, a highly sensitive person. I have friends who seem to be totally unaffected. But are they really? Whatever is in our imaginations is experienced as reality for our bodies. We know how imagining something can trigger the body to accelerate the heart and stimulate other physical functions causing the release of happy-making endorphins or our unhappy-making cortisol.

Masters are those who have gained control – *by release and expression, not suppression* – over their minds and imaginations and surrendered them to their Presence to manifest directly from their unique blueprint of qualities. You will gain your own badge of mastery through the actual practice of manifestation. A powerfully focused imagination can grow a Living Vision that is released into manifestation like a potent electrified seed. A seed planted in the unified field is activated to grow by heightened positive emotions, sureness and trust. Trust is really another name for faith and like a muscle it can be developed and strengthened with determination and practice. When we are in full alignment we can issue an almighty powerful decree, planting our seed in the unified field. We water it with heightened emotion, passion for the outcome and we leave it to germinate in the warm rays of the Presence.

Leaving it to the Presence

I really wanted to take my daughter on a trip to Bali. It sounded like the trip of a lifetime, one that would potentially become an amazing memory for us always to share and a transformational experience. The problem was the trip was very expensive. We were living in London and the cost of flights alone was daunting, not to mention the costs of the tour, food and accommodation. But I really wanted to take her and felt a strong inner guidance to do so. I dreamed of the trip and fueled my vision of it with love and saw myself there with my daughter very clearly and let it go. A few weeks later all my jewellery was stolen. I had worked as a jewellery designer and also loved jewellery and had a beautiful collection of unique pieces. Everything had been stolen except the very sentimental things I had been gifted. It was a struggle with the insurers but an amount finally came through – the exact amount to the penny that we needed to make the trip which did end up being an absolutely incredible experience!

How the money came to me did sting a bit, but experiences are worth more to me than things. Perhaps I could have been a little more specific in my Creative Alchemy?

In Bali it was very difficult to get phone calls. Someone had to cross a field to find us to bring us to the office phone if my husband called at the place we were staying. I missed him and wanted to speak to him often just to touch base. I worked out the time difference and when he would be just waking up, I would send a mental eagle to nip his toes. Invariably he would call within ten minutes. With energy, there really is no distance – and thought is energy. It travels on electromagnetic waves just like sound. Directed thought becomes an arrow flying to its goal. Of course, there was no problem drumming up the requisite amount of heightened emotion for the job!

This story illustrates a few things, one of them apparent in the way I manifested my desired outcome of the trip to Bali. I managed to manifest according to my vision, however, afterwards I needed to dial backwards to work out why my results had to manifest in such a seemingly destructive way, the theft of my jewellery. This is subject to personal interpretation.

What became clear to me was that this was the largest of several small break-ins I had experienced that year. I was living above a pub and theater I helped to run and so the whole world and its wife were underfoot, so to speak, every night. I was also completely sacrificing my life path for my husband's life path. Things looked good from one perspective but from the perspective of my authentic expression of a meaningful life, less good. I had decided that his life's path was more important than mine and out of love was supporting his passion by making it my own. I had put aside my dreams and my own career.

By analysing the break-ins and thefts and asking myself what the opposite scenario would be I could tease out the shadows playing out in my psyche. I discerned that I was not *receiving* and needed to break out. It perhaps seems overly simplistic, but it was accurate. When I started following my own star, the break-ins stopped.

You may experience the same sort of thing when you stay in a relationship, job, or place past the time when it is a growing, nourishing experience for you. Sometimes when the proverbial starts to hit the fan in a big way it's a message it's time to move on and you're just not listening.

Don't take it personally, take it as a clue on the treasure hunt. Courage is our friend in this game.

The Visionary Third Eye

The pineal gland, also known as the third eye, is shaped like a tiny pine cone tucked away in the middle of the brain. It secretes melatonin, a serotonin-derived hormone which regulates sleep in both circadian and seasonal patterns. It is also one of the most important ingredients in our personal transformation and fuels the crucially important ability to visualize. You could look upon the pineal gland as the projector of dreams on the screen of life or as the conscious creator within the body. The pineal gland has been depicted in art since the beginning of civilization and can be found in the ancient cultures of Indonesia, Babylonia, Greece, Asia and Central America (Mayan, Aztec, Incan and Olmec). It is rendered in the hieroglyphics of ancient Egypt, in the large Baroque sculptures of saints in the colonnades outside St Peter's Basilica in Vatican City, in the bas-reliefs of gods from the ruins of Sumer and the artefacts of many other cultures. Even the all-seeing eye on the American dollar signifies this gland and its ability to perceive subtle realties.

The illusion of separation dissolves when the third eye is cultivated. The characteristics of this gland and the effects of its hormonal secretions include clarity, concentration, intuition and insight, decisiveness, the perception of subtle dimensions including energy fields and movements of energy, inspirational creativity and bliss. And visionary experience! It's the captain of your imagination and the vehicle by which you may apprehend dreams beyond your own. The pineal gland is stimulated by sunlight[119] and being kissed on the forehead! Both are expressions of love. It's always advisable to connect with your heart before activating the pineal gland to ensure you are in alignment with your highest good and the highest good of all.

DMT (N, N-Dimethyltryptamine) is a psychoactive chemical released by the pineal gland at birth, death and while dreaming and seems to have a role in consciousness. In the pharmacopoeia of our bodies, DMT may be responsible for altered states and perhaps even the magic flying carpet which transports us to 'spiritual realms' of extraordinary joy and timelessness more real than real.[120] It has been named the 'spirit molecule' by medical doctor, psychiatrist and researcher Rick Strassman.[121] The psychedelic visionary effects of psychotropic plant medicines such as ayahuasca are caused by DMT in the plants themselves.

It's definitely not necessary to ingest DMT to experience transcendent states of being. Just bringing ourselves into a truly relaxed state can be enough. Awareness, relaxation and intention are the main tools for activating the pineal gland. Studies have demonstrated chemicals and cascades of hormones are released during the intense relaxation characterized by a Theta brainwave state that have substantial benefit for physical and emotional health.

Some of these hormones are endogenous cannabinoids responsible for 'exercise euphoria' (*endogenous* means 'originating within the organism'), melatonin (which can stimulate dreams) and also our own natural supply of DMT. This research provides a physiological understanding of emotions and experiences that Western medicine has previously been unable to explain. [122]

The pineal gland also converts light, temperature and magnetic environmental information into neuroendocrine signals, which regulate and coordinate physiological functions. Sun gazing is one of the most powerful awareness techniques as it stimulates this gland. Many cultures, including Mayan, Egyptian, Aztec, Tibetan and Indian, have at one time or another used this technique, believing it helped heal illness and imparted superpowers, such as extra sensory perception and the ability to live on *prana,* the life force. Forty-five minutes after sunrise and before sunset the sun does not release ultraviolet (UV) rays which can harm our eyes. The UV index should be zero at this time, but you can check to make sure and there are several apps for monitoring UV index available. You can stare into the sun during these UV free times, ideally while standing barefoot on the earth or the grass for a few seconds – adding ten seconds a day each day up to 30 minutes. This is a significant immune-system booster and also strengthens the pineal gland.

NASA scientists began to call the practice of sun-gazing the 'HRM phenomenon' after submitting 'sun eater' Hira Ratan Manek to testing for over one hundred days to confirm that he was living on sunlight, also noting that his pineal gland was enlarged. [123] He ingested no food (or water). The point of using this technique is not to become a sun eater but to charge your brain and body and vitalize your pineal gland. Practicing visualization also stimulates this gland.

Studies have revealed that sodium fluoride, bromide and chlorine are absorbed by the pineal gland and calcify it, negatively affecting our ability to dream and vision. Ingesting these chemicals should be minimized. When fluoride accumulates in the pineal gland it formulates phosphate crystals that cause the gland to harden. [124] This interferes with the manufacture of melatonin, a neurotransmitter that is important for regulating our sleep cycles. [125] If you use sodium fluoride in your toothpaste, do not swallow. Filter your water or access pure spring water if possible. Mercury, pesticides, alcohol, tobacco, caffeine and refined food can also calcify the pineal gland. I do love my coffee. Moderation is key.

When the pituitary gland and the pineal gland are fully activated the hemispheres of the brain function in harmony. In Hinduism they would say that the dance of Shiva and Shakti begins. Shiva represents the primordial cosmic power and Shakti the creative power. When these forces come into balance, an alchemical wedding or inner marriage takes place and all aspects of our being come into alignment and coherence. We achieve inner sovereignty. We are no longer masters over others nor are we their slaves. Our inner vision is fully activated. We see beyond the 'veil' to other finer realities and can draw finer energy into our day-to-day world to create positive change and inspire others. Our ability to imagine and visualize increases.

The third eye is the gatekeeper of senses beyond the main five and the portal to transcendent experience. It is our agent of seeing in the imaginal realms where we may meet our guides and our own Presence. Our senses can intuitively perceive these guides. With time, they can be heard and sometimes seen. Some of these guides are ascended masters, so-called because they have completed Earth university and now exist in a finer higher frequency body visible only if we increase our own frequency or they elect to step down theirs. They are 'prisoners of love' sacrificing their own evolution to take us by the hand and pull us to freedom. Some of us have been recycling the same patterns for lifetimes.

My story of connecting to their guidance is in my memoir, *God's Theory of Creativity: an odyssey,* a kind of 'warts and all' blow-by-blow record of radical awakening and multidimensional intervention. Earth university is a free will curriculum and we need to ask for help for them to intervene. They help us to attain expanded consciousness, achieve greater wisdom, deeper love and stronger, more focused power. They oversee us when we decree to strengthen our connection to our Presence and add their own specific essence and qualities if asked. View them as our elder brothers and sisters. Their advice is often very practical. It's all very orderly! We just need to turn our antennae to the right station.

The experience described below happened 11 years after I began to receive this direct guidance, around 1995. I was in a very relaxed transcendent state, which is the key. I had experienced dream visions and inner visions as a child and off and on as an adult. I had experienced seeing energy and light emanations and knew how to sense the Presence of these beloved beings who are here for us all. But I never expected to see with my two physical eyes in this way.

I Danced with Christ in Bali

I was on my own this time, my second time in Bali, at a Buddhist retreat. It was an exceptionally hard time. My husband had died and I needed to heal my soul or I wouldn't make the completion of my Earth journey. Burgs was a Buddhist master I had met in London with some friends of my daughter. I had loved his deep wisdom coupled with a cheeky irreverence. Very few of us were there in this exquisite place on the quiet, non-touristy east side of the country. Burgs was now back in England and it was just George, the teacher assistant, a young man being treated for a serious illness and the people who ran the place and me.

The energy was very heightened and otherworldly. Bali is a place where the veil between worlds is very thin. Every night I would walk down to the shore in the velvet blackness underneath the most brilliant pulsing canopy of stars I had ever seen. There was a huge statue of Buddha and I would sit in his lap in a state of rapturous love hugging him and calling him 'Papa' as if he was really there.

One day I was practicing Qigong with George on the wooden deck. The deck was wide and open to the turquoise sky and the cerulean sea. There was a feeling of floating, endless blue above, endless blue below. George had put on some really exquisite, sensual, floaty music. He had led me through some exercises and we were now flowing with the energy inside and outside us. George was deep in his own practice. There was an overwhelming stillness. Beautiful music, soft Balinese air, endless blue, synchronistic flow. I was in the zone and pretty blissed out. I formed a large energy ball, as he had instructed me to do and with two hands pressed it into my heart.

In that moment Christ, Kwan Yin and Buddha appeared before me. Christ was to the fore, Kwan Yin and Buddha a step behind. I was not seeing them with my inner vision but with my normal eyes. They were right before me, surrounded by the deepest most intense shimmering green light. Tears began to stream from my eyes. George, probably thinking I was having some sort of cathartic emotional release, went and sat at the edge of the deck but I wasn't aware of him doing so until later when I saw him sitting there swinging his legs and looking out to sea.

Christ came forward and bowed his head. He took my hand. I curtsied deeply and took his hand and we began to dance. It was as if I had danced with him in this almost medieval, courtly way a thousand times before. I was completely present in the experience, there was no inner witness or self-consciousness and therefore it was neither remarkable nor unremarkable. I was fully present and immersed. There was a deeply ecstatic edge to what was happening. My heart had swallowed me whole. He became my heart. My heart became his.

It could have been an eternity or a minute. Then it was over, they were no longer there, or in any case no longer visible. There was the blue sky and sea, tears still flooding from my eyes and George swinging his legs, trying to appear nonchalant.

When I have shared this story a few times with friends in the past, they have been amazed that we can have such an interdimensional experience. We can. To us, ascended or cosmic beings are usually invisible because their vibrational frequency is so much faster than ours, just as the blades of a fan disappear when they are whizzing around. But if our frequency increases, such as mine did that day by virtue of where I was and what I was doing and being incredibly relaxed, there they are. And in some cases of rescue missions, ascended and cosmic beings will step down the frequency of their vibrational field to match ours. The universe is an incredible multidimensional treasure chest waiting for you to spring its lock.

Summing Up the Power of the Imagination

The imagination is our main creative tool. It is how we negotiate our reality and how we create it. To understand how to focus it and empower it into Living Vision is to become a master of our reality, a captain of our ship. There's a saying in theater, 'If it's not on the page, it's not on the stage.' This means the script is crucial to the theatrical production. The script or play of our lives is created in our imagination.

As Artificial Intelligence accelerates, gradually replacing many functions in our society, the need to develop our imagination becomes even more crucial. Our individual imagination is what makes us unique. Aside from the ability to love, it is our greatest gift. And it can be developed. Like a muscle, the more we use it, the greater its strength.

- Everyone, no matter what their lifestyle or profession, should have a creative outlet. This is not a matter of talent but of human necessity. Creative acts build the imagination, especially if they are unrestricted. Dance, painting, even just playing with colors, drawing or doodling all come under this category.

- Practice exercising your imagination. Look at a flower and imagine what it would be like to be a flower. Imagine what a flower would say if it could speak. This may seem fanciful but this is how the original medicine healers learned the properties of plants.

- If you have a dream you wish to manifest, imagine it. Imagine all the details. Imagine the dream fulfilled. What does it look like? What do you feel like with this fulfilled dream? Imagine every detail. Write it down. Draw it. Learning how to visualize is very important. Don't worry if you can't now. There are techniques in the practical application to help you.

- Imagine different possibilities for your life. What do they look like? Feel like?

- Imagine yourself living in a different era, or as a known historical figure. Imagine what your life would be like. What are your clothes, your surroundings, the smells and sights around you? It's not a matter of right or wrong, it's an exercise to increase the flow of imagination.

- If you are undergoing stress, imagine an idealized version of your current situation. This is not escapism. This is creative thinking. You are literally imprinting the energy fields with a new template and changing your neural pathways and other biological functions. Charge that vision with love and hope and excitement.

- When you feel bogged down by the news, imagine the Earth the way you want it to be. This is not a futile exercise. We live in a unified field. Who knows who you might be giving hope to in that moment! We can solve the problems of our current reality while dreaming a new one. It's a science, not a diversion. This is a wonderful exercise for children.

- Imagine a circle of protective light as bright as the sun around yourself, your loved ones and all beings everywhere.

- Call upon your guides. Center yourself, breath evenly and let them know you are ready for your next step. Grab your journal and begin a written dialogue. Ask questions, say how you feel, seek solutions to problems. Then write back. One day there will be an answer you didn't expect and you couldn't have known and another wonderful journey begins.

5

THROUGH THE LOOKING GLASS – THE POWER OF IDENTITY

Thanks to impermanence everything is possible.
– Thich Nhat Hanh

Chapter Five will examine identity and see how ideas of who and what we are, absorbed by us from culture, gender, family, school, race, country and so on, can influence how we perceive the world and our beliefs and therefore our destiny. For example, is the identity you have unconsciously chosen to wear keeping you in less than ideal financial straits? Or in a job you don't love? When you understand how identity is constructed it becomes easier to change your identity and finally it becomes as easy as changing clothes.

One of the purposes of Creative Alchemy is to empty your mental and physical house of any traumatic emotional memory that is suppressed and buried there which then allows the architecture of your beliefs and perceptions to be restructured according to the reality you wish to inhabit, uncolored by regrets from the past or anxiety of the future. As the accretion of thoughts, memories and emotional reactions begin to drop away, your beliefs about yourself and the world start to shift and ideas of what makes you who you are dissolve. Your old identity begins to fall away like an old husk and a new, fresher you will emerge. This higher iteration is less constricted identity and closer to your true essence.

This process causes the flow and reach of your consciousness to increase. As the flow of consciousness increases, you become a more evolved iteration of yourself, the self you aspire to be. When more of us come to understand this alchemical process, we will be able to solve our planetary problems and co-create a new world. The prophesied Golden Age of Peace on Earth? A prophecy is a probability and with intention, will, passion, vision and action it can manifest.

As the false or 'borrowed' identity falls away we see life differently. It's no longer about success or failure, old-paradigm concepts. It's about learning. How many times did Edison 'fail' at designing a workable light bulb? Over one thousand times according to legend. In his mind, he had never failed but the creation of the light bulb took one thousand steps. [126] We don't give up. We don't believe 'outer appearances' as anything other than clues to our internal states of being. We know that, if we can imagine it, we can manifest it, with the right steps. There are no failures, only learning experiences.

We no longer need to hustle. *Our perfect expression of life becomes a consequence of being.* Self-criticism falls away with the constructed self and with it the tension of wearing a mask. Our constructed self is the 'made-up self.' It's like an intellectual suit of clothes made up of concepts we adapt for our own purposes or that were imposed on us as children or students by family and culture. We come to believe this clothing is who we are. It becomes a fixed identity, a rigid armor that interferes with fluidly moving forward in life. As the constructed self falls away, we disrobe our minds and become freer and closer to living aligned with our true essence. Here's a peek at some aspects of our identity we can address to reach the naked self.

Let's begin with the much-maligned *ego*, home to our identity. In this great era of integration, it is time to integrate all aspects of our psyches.

Integrating the Psyche: The Path of Illumination

Currently many of the paths to attain an illumined, liberated state call for excising or banishing the human ego. Creative Alchemy does not. The term *ego* is bandied about continually and often used as a criticism flung at those we feel are behaving inappropriately. What exactly does it refer to? Does it have a physical home or is it an abstract psychological construction?

Here is a little history. The word comes from Greek and means 'I am' in the sense of self-identity. Perhaps we could say it stands at an opposite pole to the I AM Presence, the great undivided Self. We could call it the *little self.* It is the first-person singular of the verb 'to be' and is a term representing the part of the self that distinguishes and separates each of us as an individual from others.

Sigmund Freud, one of the founders of the field of psychology, defined the ego as the psychic apparatus that experiences and reacts to the outside world, mediating between the pleasure-seeking and primitive drives of the *id,* our instinctual and biological component, and the demands of our social environment presided over by the *superego,* the ethical component of the personality. [127] Freud's contemporary Carl Jung felt the ego to be the hub of consciousness that forms and holds all unrepressed perceptions, thoughts, feelings and memories.[128] Nineteenth-century philosopher Friedrich Nietzsche, who influenced both Freud and Jung, felt the ego was a pragmatic fiction 'necessary to carry out the operations of reason.' [129]

Eastern mystical thought divides us into an egoic 'small self' and the undivided Self. The Indian sage Sai Baba famously said, 'Man minus ego equals God.' [130] The channeled teaching *A Course in Miracles* states: 'The ego is insane. In fear, it stands beyond the Everywhere, apart from All, in separation from the Infinite.' [131]

Psychological and spiritual models all seem to view the ego as central to identity, for better or worse. This identity is the construct of self that holds in its grab bag gender, culture, religion, race, family mores (the essential or characteristic customs and conventions of a society or community), profession, partner and so on. All these elements have little to do with the true essence of your eternal Self.

The spiritual model perceives the ego as the identified small self, which separates you from unbounded unity consciousness, a state wherein true, innate wisdom may be accessed along with the true reality of your being. The psychological model sees the ego as integral to your functioning, conflating individuation, time and memory, the organizer of thoughts and the collector of knowledge central to identity.

So, is the ego the insane, fearful ruler of our psyches? The ideologue? The persecutor? Is it truly the master of our function of reasoning? How would the great thinkers of the past perceive the ego in the quantum era?

Neuroscience has provided us with many new concrete ways to understand the interface between the psyche and the cranial brain. For example, we used to think that the right-brain was the creative brain and the left-brain was the analytical brain. But neuroscience has debunked this theory, saying new data reveals a whole brain model for creativity.[132] Neuroscientists have also located an area of our brain which rules our coherent sense of self – a sense that is absent or undeveloped in small children and animals and could be the physical home of the ego. This brain hub, called the *default-mode network,* goes offline when we meditate deeply, take psychedelic substances or use other techniques to trigger mystical experiences of oneness, such as yoga breathwork and running, which can cause an 'in the zone' experience. These experiences can also come upon us suddenly and also for no apparent reason, leaving us in a state of wonder, connectedness and awe, a seeming act of grace. The default-mode network also goes offline when we engage in any activity which causes us to be absolutely present, such as extreme sport. On these occasions, our constricted sense of self dissolves, giving way to transcendent experience.[133]

The part that falls away during transcendent experience, this default-mode network, is the part of our brain designed to reflect, ruminate and create mental constructions of 'self.' It thinks about thinking and reasons, among other functions, including acting as a mind manager, the brain's captain of industry (a repressive one), and a filter for external information.[134] It is also the part of the mind which immerses us in the past and the future and is given to chewing on memory or engaging in planning.[135]

Having a default-mode network is the equivalent of what happens in the story of Adam and Eve when they ate an apple from the Tree of Knowledge.

It gives us self-consciousness. When the default-mode network is online, we are metaphorically kicked out of the Garden of Eden of our inner cosmos. Without it, we could be psychedelically tripping all the time, riding a magic carpet of multidimensional states in ecstatic oneness. Like Odysseus's sailors on the island of the lotus eaters, we would forget our goals in a haze of seductive intoxication.

There is a way to negotiate between the identified small self and the expansive undivided Self. As we clear buried emotion and create new neural pathways, we can have moments of great expansion, heightened perception and awareness and be self-aware and focused conscious creators at the same time. We achieve this blended awareness by surrendering our will and the fruits of our actions not to the small self, but to the Godself, the active principal and the true doer, our inner Source of infinite supply, the I AM Presence.

When damaged and fearful, the ego *is* a liability. A wounded ego creates a fortress around us – this is a mental prison of ceaseless thoughts, accusations, ruminations, blame and projection. It immerses us in a toxic stew of detrimental neurochemicals and hormones, which can damage and addict us. It can create an armor of our musculature or bulk us out with unhealthy protective padding. Your wounded ego will suffocate you, and in the case of addictions, even kill you to 'keep you safe.'

By contrast, a healthy ego is an asset which can be used to negotiate the world and successfully achieve and manifest in the world of form. As we free ourselves of trauma, we no longer experience the amygdala hijack that causes irrational and sometimes dangerously overblown reactions to life. As we connect with our true essence we drop our borrowed social identity. As we clear out the basement of our murky subconscious full of buried emotions, that which does not serve us is alchemized and consumed. As we surrender to our Presence we become a truly cohesive whole – the universe in concerted action. Freud's tripartite psyche becomes obsolete.

As integrated beings, ideally, we receive inspiration and illumination from the Presence and the ego's role becomes the servant. It can relax. It's not in charge but it does have a job to do. That's the right relationship. The ego doesn't need to be excised; it needs to be integrated so that its functions can be relegated to tasks for which it is best suited, such as regulation. We can use all parts of our psyches, including the ego, so long as we are able to perceive them clearly and empower them to fill a useful function which enhances the whole.

A problem we have on a planetary level is that when intuition and inspiration are blocked the ego goes on formulating ideas, replicating meaningless architectures of thought that then become meaningless things, continually reinforcing a dysfunctional status quo. The ego does not like change even if it is beneficial. Sound familiar? When we have techniques to increase inspiration, such as those that Creative Alchemy makes available to you, the ego stops being so noisy and intrusive. We are the ones who put it in the driver's seat. We can reverse it.

The ego is not a barrier to truth that must be banished – an aim that only leads us to play out our old games of rejection and abandonment. If we try to exile it, the ego will struggle against our efforts, emerging often more monstrously and controlling than before. It is an aspect of us that needs love and healing. Our greatest strength can be found in the healing of our deepest wounds. When the ego is out of control it is like a wounded teenager acting out. When it is healed, it becomes the mature administrator of operations. The faithful employee of your magnificent Higher Mind. When we express and heal our wounds through the techniques of Creative Alchemy, we become free and it transforms in nature from lead to gold. That is the Great Work. The ego gradually becomes something of infinite value. It becomes a noble assistant which can help us build a beautiful life.

Eventually, as an inevitable process of alchemy, your ego is subsumed entirely into your Higher Mind. This is when you have totally surrendered to your higher purpose and higher path, a process that needn't and shouldn't be forced. I personally learned the meaning of the idiom *you can't storm the gates of heaven* by forcing things and having them blow up in my face! Accepting and allowing are keys to transmutation of the ego.

The Ego is the Horse That We Ride

When my daughter was 12 I took her to Bali where we had incredible experiences. This was one of my earliest trips to this very special place I've been drawn to several times. Our hosts were Madé Surya and his American wife, Judy Slattum.

Madé and Judy took us to many sacred sites where we saw the hidden Bali and learned a great deal about its mystical traditions. We made special visits to meet some of its Balians – traditional Balinese healers who are chosen by Spirit, often resisting fiercely, even hiding, until finally succumbing to the great force that has singled them out. We learned about the unusual religion based on Hinduism and Buddhism combined which imbued the atmosphere with mystery and beauty. There was a very spiritual feeling among these people who have an understanding of energy and dimensions which far exceeds ours in the West. They are in the flow and a sense of grace and ease permeates the atmosphere.

Madé's father was a Hindu priest. When rats took over the family's crops his father went into a deep meditation and summoned the Rat King, the Deva (archetypal intelligence) of that colony of rats. He offered the rats one strip of his land. *This was accepted and they duly moved there, so the possibility of plague was contained by nonviolent means. This communication between dimensions was very commonplace in Bali.*

We learned many things from the Balinese, and Madé's view of life. Pertinent to the examination of the ego, he told us that he saw the ego as a stallion. If we controlled it and rode it, it could carry us to victory in our lives. If we let it get out of control, we would land in a ditch. I love this description and think it creates the right understanding and relationship and I believe it set me on the path to my current understanding. Thank you, Madé Surya!

Embracing Your Shadow

The shadow is the part of yourself that you reject or project. It is everything you think you are not. It is everything you fear and don't want to be. It can also be everything you long to be. It is the seeming opposite of your outward identity. And it is very powerful. We often project the shadow onto the world and attract people who represent these suppressed and disavowed parts of ourselves. Its signature is attraction and/or aversion and it pulls us off center. We can identify our shadow by what we are obsessively attracted to or repulsed by.

Often people obsessively seeking the light are suppressing their shadows. A kind of tyranny emerges. Ideas like *'We must be happy and light-filled all the time,'* as well as a long list of other *musts* and a lack of acceptance of anyone who varies from these rules is imposed on us by our repression of the shadow. Life becomes a series of platitudes and rigid dos and don'ts.

Think of the strong reactions about money some people have, either defining their worth by it or completely shying away from it, or about the lack of compassion and even anger many demonstrate towards those who don't share the same values.

This phenomenon can have more alarming and tragic consequences when the projection of shadow onto whole groups of people leads to persecution, such as Hitler's persecution of the Jewish people or the persecution of African Americans in the United States, classic projections of the shadow that spread like wildfire, linking shadow to shadow across the lands.

To work with the shadow, we need to be honest and authentic. We need to look into our psyche and lovingly acknowledge what is there. Often it's not pretty. It could be narcissism, a kind of self-obsession, jealousies we won't admit or insecurities and fears that cause us to compare and judge and resent. Shining a light into the shadow is an important step in the alchemical process of turning what you find there from lead to gold. Narcissism can be transmuted to healthy self-love. Jealousy can be transmuted to healthy self-acceptance and a feeling of being enough. A lust for vengeance can be transmuted into a quest for justice. There is often brilliant treasure hidden in the shadow. Sometimes our greatest strength is hiding within the transformation of what we perceive to be our greatest weakness.

Here's an example of the fruits of shadow work.

Dancing with the Shadow

Beth's childhood was traumatic. She had experienced abuse and alienation and had been rejected by her parents for reasons that were never explained, forced to live with distant relatives at the age of nine. Her experiences of power in her new household were very negative, creating her belief that all power was bad. Her perceptions were influenced by extremely traumatic buried emotions. She believed the world to be a dangerous place.

In adulthood Beth developed an identity that was loving and kind, very spiritual, highly principled and easily offended. She was against many things, especially abuses of power but really anything that represented power in an upfront way. Her main trajectory was to get more advanced spiritually, to become a living angel. This goal was creating a lot of internal pressure for her and generating ruthless self-criticism. Creating community art projects was her profession.

So why was Beth's life full of bullies? She experienced the destruction of many of her projects by unreliable and even Machiavellian partners, who seemed to take pleasure in destroying whatever she tried to create. Her romantic partnerships were also painful, fraught and dysfunctional, with her usually being cast as the victim being persecuted by another.

Using the techniques of Creative Alchemy, Beth was able to access many suppressed memories and release the pain of her buried emotions cathartically. She was diligent and continued the process with committed and sincere application, observing her reality and checking back through her beliefs and perceptions to find yet another buried trauma and readjusting as needed. She began to look at finding hidden emotional memories as a treasure hunt. In the process, she confronted her shadow and discovered there were jewels to be had there. She was much stronger than she thought.

As she cleared the clouds from the lens of her perceptions, she began to see how all the people bullying her were reflecting back to her a suppressed aspect of her emotional landscape. She didn't take authority, she passed it on to others. She shied away from any show of power, fearing its destructive qualities and denying its constructive qualities. She saw she was living out a kind of spiritual bypass, meaning that she was ignoring all the information her body was trying to give her when intense feelings emerged. This information had to be reclaimed from her shadow and incorporated with the rest of her psyche so that she could develop discernment.

Beth is now more empowered. She runs a successful company that creates theatrical presentations for communities. She is in control, leads a meaningful life, wields a gentle power and is surrounded by helpful people. All this is possible because she applied the creative alchemical process diligently.

Shadow work entails examining your reality to discern what triggers you and using your triggers like a treasure map to discover what traumatic memories and emotions you have suppressed. When we are extremely polarized inside ourselves, what we feel strongly opposed to or strongly attracted to will often shine a light of truth on what we have suppressed in our inner cave of secrets.

In the West many people reject spirituality and embrace science and then make a religion of science and defend it to the death, acknowledging no deficits. Others reject science and feel anything emerging from the world of science and conventional medicine is evil. Rejecting one or the other would impoverish us.

When our shadows come to visit, we want to explore their jewels. Often what we have most vehemently rejected are the wounded parts of us that we have polarized and suppressed – and they hold important meaning for us. If you continually draw a type of person into your life who wreaks havoc in your existence, is there something in their personality you are not seeing in yourself? The victim always draws the bully to it and the bully always draws the victim. It's the law of the shadow.

Try not to over-identify with a side in any pair of opposites. If you're against something, try to see the other point of view. Build the new rather than protest against the old. Transmute rather than deny. Don't do yourself an injustice by dismissing whole demographics of people because they don't share your worldview. Examine what you are against very carefully and stop judging, for you will inevitably draw judgment to yourself if you do. This is the Law of the Circle. It is possible to take action to correct destructive situations whether they are political, social, environmental or personal without targeting individuals with projected emotions.

We must challenge ourselves. We can learn to integrate an opposite *or* hold two polarized ideas simultaneously – the divine paradox that melts duality. What if we approached our work as play? Can we love money *and* meditation? Can we be spiritual and also wealthy? Can we find the similarities in science and spirituality, both noble quests for truth? Quantum theory is certainly a bridge. Can we embrace power and be a force for good? Clinging to one pole in a world filled with duality draws us away from change and evolution.

As an exercise, look carefully at everything you have an extreme aversion or attraction to and then look at its opposite and examine it against the reality you currently inhabit. Are there qualities in the hate list that you believe are missing from your life? Very likely. Try to do a few things that are on your aversion list and begin to release the potency of this rejection. But only so long as they are not harmful to you or someone else.

However, you don't need to go as far as the Mad Monk!

The Mad Monk of Bhutan

Drukpa Kunley (1455–1529) was known as the Mad Monk, the Divine Madman and the Mad Saint of the Kingdom of Bhutan. If you visit his monastery you will be blessed by a foot-long wooden phallus evidently modelled on his own. The image of this same phallus is painted on many of the charming homes of the Bhutanese who revere his memory. Some of these have big eyes painted on mushroom-shaped phalluses, merrily spurting, a sign of abundance, with ribbons tied around them.

The Divine Madman achieved enlightenment by embracing his shadow. He did the opposite of everything he was expected to do and wholeheartedly embraced his aversions. He taught Buddhism with singing, crazy humor and outrageous behavior. He admired women and taught them via his 'flaming thunderbolt of wisdom.' He provoked, poked and jolted his devotees with obscene behavior and humor and broke the rigidity of orthodoxy he felt was strangling the true tenets of Buddhism. In doing so he ultimately achieved full liberation and unity consciousness.

The Mirror Game

Life is like a hall of mirrors. There is usually a sobering moment on your journey of illumination where you decide to figure out why there is a sticky repetitive pattern in your life. Ending the pattern starts with the mirror game, a good tool for many of life's situations with regards to the treasure hunt of suppressed emotion. This is where you admit to yourself that whatever someone (or life) is doing to you, you are doing to yourself. It's 'next-door' to shadow work, with a slightly different emphasis.

Commit to looking at every single thing in your life as a reflection in the mirror of your subconscious and your conscious self. You don't need to dig deep to get results. You just apply it to everything you experience in your day-to-day life as a reflection of your inner state of being.

Everything loving? People kind, helpful and considerate? That's a healthy reading.

If everyone is angry around you and you are a quiet, kind sort who never gets angry, time to check if you have a lot of anger inside you that is very suppressed. It's likely taking all your power and energy to keep that anger suppressed. But you won't get away with suppressing it forever because the mirror will continue to show you what's going on by drawing an external experience that matches your inner state. If you're exploding with irritation, annoyance and barely suppressed anger you will find either angry responses or fearful colleagues, partners and friends. What are you afraid of that your anger is masking? Anger is a signal we feel our boundaries are being transgressed.

Or, you are a big giver and quietly seethe with resentment that you are underappreciated, that no one gives back. Ask yourself this: How good are you at receiving? Not too good? How do you treat yourself? Not too well? The mirror tells the truth that we haven't admitted.

You'll know you've unraveled your stuff when the world starts to be consistently nice to you and abundance and helpful people flow easily to you and your relationships start being fun. But you can't unravel your stuff until you know what it's packaged in.

Our inner feeling world, consciously or unconsciously, creates patterns in the fields of energy all around us. These patterns literally rearrange the manifestation of 'reality' coalescing around us, mirroring our internal world and magnetizing situations that are a frequency match.

Here's an example of inner reality affecting the outer.

Honor Your Dream

Jake took over running his late father's business as he was expected to do. His family was thrilled and made him into a kind of hero for stepping away from his dream of being a builder of sailboats. He had been a keen sailor all his life and was apprenticing with a master builder at the time of his father's death. Jake gained instant status and wealth in his new job but he was suffocating inside. However, he ploughed on and gave it his all, exceeding expectations in the first two years. Outside he was excelling, inside he was a mess.

Problems out of Jake's control began to spring up all around him. The local business council challenged his license. Employees left or became disgruntled. A rival company stole a copyright, triggering a lawsuit. Jake's misery was reflected in the mirror all around him, but he still had not made the connection between his outer and inner issues.

Jake's Creative Alchemy clearing process was effective in shedding light on the inner issues. By playing the mirror game, he was able to see and understand that the model of reality he was currently operating in was so at odds with the reality he desired, that he had to risk disappointing his family. His emotional release provided rich realizations of how deeply his perceptions had been affected by his family's belief structure and how much anger and resentment he held for allowing himself to be marshalled away from dreams again and again. After he came clean, the business was sold.

Jake now builds boats and holds tours for small groups in exotic places and is doing very well. By the way, his family is fine, as is often the case in situations like his. Either way, he did what he had to do and created the destiny he desired, supported by the process of shifting to his witness to gain a clear and unencumbered overview of what he needed to do and clearing some sticky emotion.

We often change or drop our dreams to please people. They may suffer a little while if you don't fulfil their expectations, but you will suffer a lifetime if you do.

Our Costumes

We often slip into personas we then find hard to shed. This is partly because the identity with the mass of beliefs that underlie it, creates a reality that we come to wholeheartedly believe in. When we are fixed in an identity it is very difficult to alter our destiny. A new destiny requires a new identity. Finally, we let go of identity altogether and become a free-floating unit of consciousness, shape-shifting as we see fit, manifesting the destiny we wish for in any given moment. We become consciousness in a body observing the world. When we can observe with the eyes of love, appreciation and gratitude, we entirely forget to concentrate on our external selves. Anxiety, the modern malaise, decreases or vanishes altogether. We begin to trust the flow of life. We move from the incessant measuring of ourselves, comparing and judging, to pure experience, as a child does. We are living from our true unadulterated essence and moving closer to our heart's desire and our dreams.

Three common overarching personas or identities that people gravitate towards are the victim, the perpetrator and the helper. We tend to have pity for the victim, abhor the perpetrator and admire the long-suffering helper. While studying these typical personality aspects at Penninghame House in Scotland, participants divided into groups to roleplay these three types. Of course, none who identified with the victim persona felt they could portray the perpetrator. But in fact, it turned out that we all could step into each role with surprising ease. What we learned from the exercise is that the helper is often an enabler, and most interestingly, the victim holds all the power, drawing the focus and attention of both the helper and the perpetrator. The perpetrator is the aggressor, the bully, the 'taker' but shares many of the same underlying insecurities as the victim. All three types are unconsciously and perhaps consciously manipulative.

We live in a world that expects us to have a very carefully crafted identity. Social media and often our career choices demand we brand ourselves. We self-objectify, looking *at* ourselves superficially rather than *within* ourselves, continually creating a kind of commodity of ourselves by adjusting the surface of who we are to fit expectations and fulfil the status quo. We may believe we are expressing and sharing our essence (and to some extent we may be), but the degree to which we try to control others' perceptions of us equals the degree of unease we will experience. New paradigm marketing is more about sharing our offering and building trust that what we have on offer is authentic and of value.

As you heal your suppressed emotions and change your inner tapes you can fall in love with your true essence, your own unique self. Some people in your life may fall away but others will come who love and respect you and appreciate your authenticity. The masks can fall away. Phew, it feels good.

> *We hammer wood for a house,*
> *but it is the inner space that makes it liveable.*
> *We work with being, but non-being is what we use.*
> – Lao Tsu, author of the Tao Te Ching and founder of Taoism

Character is destiny. Building character changes the outcome of your life in a positive way. Building strength of character – courage, compassion, stamina, discipline, tolerance and other important qualities, is more rewarding and fruitful than building identity because, remember, identity is a suit of clothes. What identity are you wearing today, tomorrow and yesterday? The hero? The seductress? The outlaw? The jester? The eternal mother? The king? The sage? The clown? The boss? The lonely outcast? The misfit? Be aware and you can have fun with it. Learning to laugh at ourselves is a great gift.

When your emotions are free-flowing you become more relaxed. The opinions of others don't have the same degree of impact. You can explore what and who you want to express. You could become very neutral. Or you may choose to be more theatrical. You could decide what identity you want to wear each morning after you wake up – change personas as often as your socks. Once you are aligned with your Presence, your true essence and unique qualities will radiate from whatever costume you choose. When you are increasingly anchored in your true essence you can play with identity without destabilizing yourself. Or you can be present as the eternal Self, free of identity entirely.

Love says I am everything. Wisdom says I am nothing.
Between the two my life flows.
– Sri Nisargadatta Maharaj

Meditation is an antidote to strongly fixed identity. In deep meditation, our default-mode network goes offline and our consciousness expands. We see deep within ourselves and expand into a broader consciousness simultaneously. We forget our personalities. Movement and Holotropic Breathwork are also really helpful to release the rigid muscular armoring that builds with emotional suppression, fixing our physical identity.

Holotropic Breathwork was created by psychiatrist Stanislav Grof MD, and transpersonal therapist Christina Grof.[136] Their research on the effects of LSD led them to study altered states and how these could empower people. We now know the neuroscience. Altered states and expanded consciousness trigger neuroplasticity in the brain that can be very helpful with trauma. They created Holotropic Breathwork so people could have the same physiological experience as psychedelics just by breathing. Holotropic Breathwork is incredibly powerful and best done with an experienced practitioner and not if you have a history of panic attacks, psychosis, glaucoma or cardiovascular disease.

Leonard Orr created Rebirthing-breathwork in the early 1960s and it became something of a counterculture movement in the 1970s and 1980s. This technique is also called Intuitive Energy Breathing or Conscious Energy Breathing. This is a very powerful healing modality. It potentially takes you back to your birth and after a session you are rebirthed into a new consciousness. Orr discovered in his early sessions that certain environments such as the bathtub and certain types of breath brought people back to prenatal and birth psychological states. There would be significant trauma release and ongoing benefits. When this type of breath gets going, you are taken over by it and your breath starts breathing you. Catharsis is inevitable.

I trained in this modality and also had many sessions personally. It seems the breath itself is teaching the art of breathing. I was able to release a lot of trauma and feel the grip of imposed identity really begin to slip away, offering me a freedom I hadn't previously experienced.[137] Breathwork is a powerful companion on the journey through life.

Identity Issues

Charlie's dream was to be a filmmaker. He had made some short films that showed
promise and had a very large technical armory based on his corporate tech job.

However, he managed to alienate everyone around him through his outward show of unpleasant self-aggrandizement that was compensation for his secret lack of self-worth. One after another his colleagues and helpers fell away. To Charlie, there was something wrong with all of them – a reason they hadn't been up to snuff and good riddance. He was in complete denial of any part he might have played in their departures. He became more insufferable, more tyrannical and surer of his big visions which underfinanced, depended entirely on the goodwill of those assisting, who were fast disappearing.

Finally, Charlie was willing to look inward. Using the tools of Creative Alchemy, he was able to unearth suppressed memories of his sometimes 'out of the blue' tyrannical father who had taken up residence in his psyche. Charlie was 'wearing' his father's psyche. The suppressed trauma and stress caused him to always hold his breath or breathe shallow, quick breaths, catapulting him into a permanent 'fight or flight' adrenaline-soaked over reactive state. His low self-worth that was a hangover from the violent criticism coupled with the exceedingly high expectations of him held by his family and his self-aggrandizement was the flip side. Charlie continued to work on his emotions and began conscious breathing. Several times in his day he would stop and ensure he was breathing all the way to his root in very slow, even exhales and inhales.

He also paused frequently to feel into his body and see what emotion or energy was presenting. His work in Creative Alchemy released grief and anger and his consciousness was gradually able to take up residence in his body, no longer afraid of what it would encounter. His whole energy field has shifted. He is a diamond in the rough coming into balance and his shine and new ways of interacting are attracting good collaborators. He is on a journey of self-discovery and embracing his own much softer persona.

He can see how he created the patterns that were plaguing him and that reciprocity and respect are necessary for successful professional relationships. He has shed his former identity and is growing closer to his authentic self. He is drawing positive opportunities and he is much kinder to himself and others.

To become someone new, we need to let go of who we are now. To change our lives, we must change ourselves. It's scary to drop an old identity but the freedom when we do so is phenomenal.

After committing ourselves to the Great Work of alchemically transmuting our psyches from lead to gold and diligently employing the ideas and techniques of Creative Alchemy, typically we don't try to carve a fixed identity. We begin to intuitively understand that rigidity will interfere with the creation of the destiny of our dreams. A fixed identity holds us in the past. We shift from the concept of identity to the concept of *becoming*. We become more interested in the qualities we wish to experience: joy, abundance, prosperity, whatever they may be, and what gifts we wish to share with the world. The looser you hold yourself in an identity the easier it is to fluidly create.

Know Thyself

Back in the early part of the 20th century, a great sage and holy man named Chellappan was driving around Sri Lanka in a horse-drawn cart when he saw a young man on the side of the road who he perceived to be close to realization. He called out to him, 'Who are you? There is not one wrong thing! It is as it is! Who knows?' In that one fateful moment, the young soul dissolved in a sea of light. He became nothing and everything all at once. We know the story because the young brahmachari (an Indian term for one on the path of Braham – the constant reality of ultimate truth) became the satguru of Sri Lanka, Yogiswami, and he told it to his followers. Yogiswami became a perfected being living in undivided truth, an illumined human.

The superficial perspective of external self-observation (as opposed to internal) plagues us with criticisms of being 'not good enough, pretty enough, handsome enough, thin enough, young enough, smart enough,' and results in us abandoning ourselves. All that goes when you shift perspective to the inner landscape, to *know thyself* as the ancient Greek philosopher Socrates advised and ask yourself the million-dollar question recommended by the enlightened Indian sage Ramana Maharshi, *Who am I?*

Be still. Look in and seek the oneness of the Presence in the sea of light that emerges when your story and your marriage to a particular identity falls away. Become as nothing, anchored in the endless bliss of non-identity and know that from that perspective *you can become anything*. Look inward to where visionary templates are formed and shape the reality you wish. Free of fixed identity, you can now gather any identity that you need to meet a goal. You are fluid, a conscious creator being in motion. Finally, centered in the Now, our path becomes a graceful inevitability.

Harmony and Discord

Harmony is a universal principle and brings us into universal flow. The great sages of the East described the three phases of this flow as creation, preservation and destruction. Destruction is not wilful destruction but the phase of the cycle when what is no longer useful falls away. It is as much a part of the flow of life as creation and preservation. From this falling away comes new birth. It is possible to experience all three phases without experiencing discord. When we act from harmony, we are in balance and not in thrall to the fixed identity's aversions and attraction. We are in alignment. We can be the still center of the storm.

One of the main purposes of the system of Creative Alchemy is to heal discord caused by suppressed emotion and restore harmony. When we can maintain inner harmony in the face of all events we will truly become masters of energy and therefore masters of our destiny. How far you wish to go with this is personal choice. Most people just want the tools for a happy, abundant, fulfilled life, a life of meaningful purpose. Some wish to go further. That has been my path. I seek total union with my Presence and try to put that before all goals. Of course, we are already in total union ad totally free. So, the goal is really to dissolve the illusion of separation. Both paths are perfect and neither should be forced. Both are possible sincerely applying the techniques of Creative Alchemy. Intention is the first major step. Motivation can be learned.

Experiencing Universal Harmony

As part of my healing journey I received Craniosacral Therapy and SomatoEmotional Release (SER) from osteopath John Upledger DO, founder of the Upledger Institute in Florida. John Upledger noticed that there was an emotional component in his clients' physical illness and created SER to address it, combining the two modalities with profound effects. The clinic saw children and adults daily experience huge shifts. John said not a day went by when they didn't experience a miracle in the release of symptoms from issues as diverse as severe PTSD for veterans to cerebral palsy in children, and more.

I spent two weeks at the Upledger Institute with ten other people being treated by six therapists who had been trained in sensing energy and energy blocks. They had all been through their own emotional clearing and had become highly attuned masters of their discipline. Upledger maintained that the shunting of the spinal fluid is even deeper than the breath and was a tool for sensing any blocks in the body and mind.

I would tell him of an issue in one area, but he would come over and 'talk to my body' and work on an entirely different area, worrying away at one spot that would finally cause a giant reflex in some other part.

I released things energetically that made a ting that the practitioners and I heard when I spat the energy into a bucket. In one session, a musician 'played' my energy field. This was a turning point for me in my alchemical practice. I had been diligent with my commitment and had moved a lot beforehand that had been stuck and was really ready for work on the very subtle levels. I was able to confirm what I had been sensing about energy fields and how they interface between the body, mind and emotions.

Finally, at the end of the two weeks of clearing, I felt there was nothing in my energy field but me and I entered into a state of homeostasis, a state of extraordinary equilibrium. I had a peak experience, a state of greatly expanded consciousness. The term was coined by psychologist Abraham Maslow, who had noted that certain experiences could lend us the perspective of the great sages, mystics and saints. He felt that peak experiences were non-ordinary, illuminating events that could be part of ordinary life and occurred with increased frequency with psychological health.[138]

I can only explain the experience by describing that in that moment, which stretched over several days, I could hold paradox, all opposites coming into balance. I felt the underlying harmony of existence and it was extraordinary. I was in the garden beyond good and evil that the Persian poet and Sufi mystic Rumi wrote about. I beheld and felt and was part of and totally encompassed by the underlying harmony of the universal field. I knew beyond the shadow of a doubt that its nature was goodness.

I learned a great deal from this experience and later did some training in Craniosacral Therapy. It amazed me that we can all learn to sense these subtle fields with practice and how much it changes our perception of ordinary reality.

Motivation and Apathy

Motivation and apathy can be inherent traits in the fixed identity we have inherited. We can help to shift apathy by recognizing it as an attitude that can be changed by seeking out its root core. This can be suppressed emotion or imposed identity or both.

Motivation is caused by the will to do, your intention and the love that sustains it. Your passion is the fuel of your motivation. Motivation gains momentum when it is approached as a practice. Momentum is a kind of gravity pulling you forwards and it is created by repeatedly placing your attention on something and taking all necessary steps in a sincere, committed and consistent way. Decree can help focus your momentum and enhance it. This momentum lays down new neural pathways, shifting the patterns of your energy fields, supporting your goal. Decrees that have been said for millennia are very helpful. You also gain the momentum of those who came before you.

The opposite of motivation is *apathy*, lack of physical engagement, lack of productivity and lack of initiative. Most of us have at some point experienced lethargy or even a sense of hopelessness that stops us from going forward and taking the risk of leaping into new experiences. Apathy is caused by surrendering to tedium, suppressing your dreams, surrendering to a life path that is not your own or suppression of unhealed emotion. It can come after a rejection or failure that has left you demoralized or it can be a response to harsh circumstances that seem immutable either presently or in the distant past. It can come from parents who have never reached for their own desired destinies or who have repeatedly pounded you with messages like, 'Don't get above yourself,' or 'Money doesn't come to the likes of us,' or 'Don't boast,' ideas forged in their own inherited identities. Apathy also can be a fear of losing control.

If you are suffering from apathy, it will be one of the first things you need to tackle if you would like to alter your destiny and find your meaningful purpose. People suffering from apathy do not give themselves what they need and become distanced from their essential selves. What do you love to do that you haven't done in a while? What hobby could perk you up enough to just get those embers flickering again? Is there a place to go that uplifts you and raises your inspiration high enough to let those flames catch? A concert, gallery, sports event, movie or nature walks? Exercise really helps.

Apathy often covers suppressed anger, grief or fear. The longer we put a lid on these emotions, the deeper they get buried. Getting moving in any way you can. Dance or walk every day. If you can't get to an exercise or yoga class, do some stretching or strength-building at home. The time it takes to create a new neural network by repeating a desired activity every single day ranges from 21 to 60 days. If you miss a day, start again. Neuroplasticity is the brain's ability to reorganize itself by forming and strengthening connections. Our brains are infinitely mouldable. But remember, practices should be ongoing because otherwise the neural network associated with them becomes dormant. Fold your practice into your life and it can be and should be a great adventure with increasing rewards.

During a study led by neuroscientists at Oxford University, decisions made while playing games were analyzed with brain imaging. These revealed that apathetic subjects were less likely to accept offers of rewards which required effort to procure (they were *insensitive* to the offers) and there was more activity in the part of the brain associated with potential movement. In layman's terms, they had *analysis paralysis*.[139]

Apathetic people use more brainpower than motivated people when contemplating effort! Interestingly, but not surprisingly, the researchers surmised this could be the cause or effect of apathy, meaning the jury's out regarding which comes first, the neural activity or the behavior that creates it. In either case, we can use our understanding of neuroplasticity to change this pattern of behavior.

The move from apathy to motivation in any area of your life will always be activated by a decision, which in turn activates will and an intention that is intensified by the level of your desire or need to change. You are reading this book because you want to change your destiny. You have a conviction. You decided. No doubt you have dreams of where or what or who you'd rather be.

Don't frighten yourself with big goals while you work on clearing buried emotions that may have created the apathy, just knock off one step a day. Gradually the wonders of neuroplasticity will rewire your brain, which will become empowered by the energy released from your previously buried emotions. We are remarkable in our ability 'to pick ourselves up, dust ourselves off and start all over again.' The Human spirit has a way of triumphing over adversity. The phoenix rises from the fire again and again, each time more refined and aware.

Abundance

The process of Creative Alchemy is a neurological, physiological, emotional, spiritual and mental transformation of consciousness that will enhance your life and eventually bring abundance to all areas of it. You can choose an abundant persona and transform an identity built on poverty consciousness by keying into this universal quality. Life is naturally abundant. Look around at nature! The truth is your abundant nature is an eternal quality and is beyond identity, but faking it until we make it is sometimes a necessity. Abundance is an energetic frequency, a vibration that attracts more of itself to itself. The more refined and clear we become emotionally and mentally, the quicker our personal electromagnetic frequency oscillates and vibrates. Like attracts like and the universal cornucopia comes pouring in. Good practices of self-care are important.

What steps are you taking to look after yourself? Self-care creates harmony and harmony is connected to the free flow of abundance and, most importantly, the sustaining of it. It also builds healthy self-worth. The abundance we all seek in so many ways – prosperity, love, inspiration, health, opportunity, information, well-being and so on, is a level of vibration we achieve when we have mastered self-love and love for others. It's the key to the treasure chest. When you reach this level, *abundance will seek you out*. You cannot avoid it!

Once we have cleared our traumatic emotional memory and shifted our perceptions and our beliefs we are no longer trying to manifest what we *think* we want! That is, manifesting from the wounded child inside who wants to be loved and seen and invited to the 'party.' We are manifesting from our true essence. Our true personal genius. Our transcendent imagination. Our inner superhero fueled by the power of our *energy in motion*, the unleashed power of our healed emotions. Synchronicities start to happen. What we need begins to manifest before we know we need it.

We are creative alchemists transmuting our Source energy into what we need and want and there is no limit to this with dedication and practice. Having access to that part of ourselves is a profound homecoming, an antidote to the existential malaise and loneliness we often feel within. When we are reunited with all aspects of our extraordinary selves, our perceptions, beliefs and reality shift. Heaven comes in a twinkling of an eye. It was always there, just hidden behind a clouded lens. Our dream of life begins to manifest slowly but surely at this juncture and it is purposeful, abundant and true.

Start playing the gratitude game. Be grateful for everything large and small. If we can summon up gratitude for absolutely everything in our lives, amazing things start to happen. Gratitude brings grace, grace is the avenue of miracles. Whatever identity you wish to experiment make sure you add this keynote.

A Word about Tithing

A simple practice that increases abundance and well-being is tithing. Tithing is an ancient practice that requires we give ten percent of our income to good causes. Many people who have achieved enormous prosperity have operated from this principle and tithed all their lives. Some of the wealthiest people to have walked the planet, including John D Rockefeller, Andrew Carnegie, Dmitri Chavkerov and Warren Buffett, were and are big tithers, tithing between ten percent and ninety percent of their incomes.

It's wise to tithe even if your income is small, as tithing can begin to break down the dam for you. You can tithe to good causes, charitable organizations that have benefited you spiritually or that support causes you believe in. If you're at rock bottom, tithe your time. When we are abundant we are part of the solution rather than part of the problem. We can be generous and share and enjoy good self-care. To tithe confirms gratitude and also increases well-being. There have been several controlled studies that prove spending money on others increases our well-being more than spending on ourselves.[140]

I often use the word *abundance* over the term *financial prosperity*. It gets my expectations out of the way. The universe may deliver what we need bypassing money. I had that experience producing and directing a theatrical presentation of *Peter Pan*.

Learning the Difference between Abundance and Prosperity

I had lovingly adapted J M Barrie's Peter Pan *using out-of-print books to restore this beautiful work to its original haunting and charming myth, and was assembling a production with no doubles. We were awash with mermaids, Lost Boys, pirates and Tinkerbelles. This production of* Peter Pan *was a homage to my late husband and very important to many of us for that reason. He had first seen a marvellous musical presentation of this special story at five years old on Broadway and it had set him on the path to his destiny. We had been searching for the song by Leonard Bernstein, 'Who Am I?' that he had loved from that Broadway production, and we had miraculously ended up with the world stage premiere of his entire lost score at our tiny theater.*

Ten days before rehearsals almost 50 percent of the funds fell through. It seemed absolutely impossible to carry on. We had a huge cast and a chamber orchestra. However, against all sane advice, we decided not to turn back and forged forwards knowing we could somehow pull it off.

It was a truly blessed experience as we seemed to be pulling our resources right out of the energy field. We became strongly united as a company. Everyone became involved with getting the show on, including the actors who fanned dry ice from backstage onto the stage while changing costumes. All the heads of department became extremely innovative. Magical ideas came. Our choreographer 'built' Wendy's house using the Lost Boys as walls, chimney and door while they sang Bernstein's charming song 'My House.' We had no flying rig, so our set designer innovated a small hydraulic lift hidden by dry ice that actors were madly fanning backstage. As Peter, Wendy, Michael and John moved their arms and bodies and the star cloth behind them came alight and the small orchestra played magical music, it really looked like they were flying.

That show was literally produced with love and remarkably had audience members returning many times to drink in the feeling of love and wonder that poured from the stage. It also garnered great reviews and transferred to a large theater in the US. I learned many important things including the huge power generated by a group sharing the same passionate, positive goal; the difference between finance and abundance and the huge power of faith and trust.

Summing Up the Power of Identity

The key to the whole identity game is to move away from fixed ideas of our-selves and move towards becoming qualities we wish to embody. We can play with identity and become more fluid about it. As Shakespeare said in *As You Like It*, 'All the world's a stage and all the men and women merely players.' The less we fixate on an idea of who we are the freer we become. The truth is you are an eternal being, one with the Absolute having a holiday in the world of time and form.

- What are the key identity points you have inherited from your family and culture?

- What are the key identity points you have inherited from your gender? Your race?

- What are the key identity points you have inherited from your profession?

- Keep going. Diet? Exercise routine? Religion?

- What are you strongly drawn to? What causes you to recoil? Make a list and examine the opposite. Are there extremes in your external reality that trigger you?

- Can you hold an inner mirror to your external reality? What does it reflect?

- Are you motivated or apathetic? If you are apathetic, what steps can you take to begin to shake it up a little? What do you love that you are not giving yourself? Sometimes apathy can set in when we continually override our inner voice – the one saying, *Go out for a walk, the sun is shining,* that we say *No* to until it finally falls silent. If you are apathetic, review your habits. Excess sugar, unhealthy food, alcohol and recreational drugs can contribute to apathy. Set yourself small daily goals. Walk more.

- Apathy can sometimes be depression. Depression can be 'anger turned inwards.' When you begin The Creative Alchemy Method™, devote yourself to the beginning exercises and this will start to dislodge sticky calcified emotion that could be slowing you down.

- Do you have something in your life that counts as service to others? Something that brings you joy?

- How's your abundance? Appreciation and gratitude help with abundance. Abundance is a frequency and when we resonate to that frequency, abundance finds us. What do you think about tithing? Is there a favorite cause you might find joy in supporting?

- Make a list of everything you are currently grateful for. Add ten more!

- What identity are you wearing today? What identity would you like to play with?

6

HUMANITY'S QUANTUM LEAP – THE POWER OF THE PRESENCE

That which permeates all, which nothing transcends and which,
like the universal space around us, fills everything completely from within and
without, that Supreme non-dual Brahman – THAT THOU ART. [141]
– I Am That: Dialogues of Sri Nisargadatta Maharaj

We are now diving into the esoteric science withheld from humanity for Millennia, revealed only to those who had, after enormous effort and study, proved themselves worthy of the great power the secret science of miracles can wield. Sincerely applying this science does increase power exponentially and it very wise to clear your emotions first, as I learned the hard way. More about that later in this chapter.

Most world religions and spiritual philosophies have terms for the state of being we can evolve to when we've shed our 'mortal coil' if we have engaged deeply in the Great Work of transmuting the lead of our souls to gold. These terms include the Ascension Body, the Adam Kadmon, Mahatma, Ātman, Supreme Personality of the Godhead and the Solar Logos, the sacred body of light or fire spoken of in many cultures. Christ called it Father. In Sufism, it is called the Most Sacred Body. In Taoism, it is called the Diamond Body and those who attain it are immortals and cloud walkers. For the Rosicrucian, it is the Diamond Body of the Temple of God. Ancient Egyptians called it the Akh, the luminous body or transfigured spirit that survived death and mingled with the gods. [142] Buddhists call it the Rainbow Body stating all who attain the Great Rainbow Body eternally retain their present life's body appearance until all beings become Buddha. [143] Creative Alchemy uses the term I AM Presence.

The new paradigm concept (or dispensation) is that our earthly bodies are quickening to the degree that we may potentially merge with this highest aspect, bringing unparalleled possibilities for the evolution of human potential, the fulfilment of our own highest destiny and the co-creation of a paradise here on Earth. This can only come about if we have truly transformed from thinking beings, the limits of which are all too apparent, to illumined beings, able to access and live by the wisdom, power and unconditional love of our Presence. Creative Alchemy is a powerful tool to aid this transition and will help ensure we cross this bridge with ease and grace. What appears to be chaos in the world around us is the birthing pains of a new human, Homo illuminatus. This term Homo illuminatus seems the most apt term to convey this next iteration of our evolution, the descent of our higher nature into physicality.

When we express our higher nature, we become divine humans. The word *divine* comes from the Latin *divinus,* meaning 'foreseeing,' and *divus,* meaning 'god.' To become divine means to access godlike superpowers, to become fledgling masters and then realized masters, to become the image of that which we emerged from, the universal Source. With our superpowers accessible, our ability to innovate and problem solve will grow exponentially. As we grow and deepen so does our ability to see the divine in everything and this invites the divine in everything to emerge.

> *The person who perceives Brahmin*
> *in everything feels everlasting joy.*
> – Bhagavad Gita 5.21

When we are connected to the Presence the world transforms. No dogma will teach us what we need to know. We need to *experience* the Presence and receive its guidance and all else will follow. Your Presence holds the highest expression of your template for manifestation, your gifts, your powers, your talents – what you have come to bring into existence to further the evolution of the Universe. This is not aggrandizement but a restoration of our true selves and our true purpose. Creative Alchemy is a road map that can be tailored to your individual needs. It is not more dogma or a call to a new set of beliefs. It is a formula that will bring you to your own profound and powerful experience.

> *Is it not written in your law I have said you are gods?*
> – John 10:34 [144]

What is the I AM Presence?

Where did this term come from? I first learned of the *I AM Presence* during a period of radical awakening in the mid-1990s while sitting in a rose garden in London with my friend and fellow explorer Christina Hagman. I had just begun to receive clear, audible guidance. Christina had always had this ability of clairaudience and was unsure of it whereas I had longed for it. Now the two of us were making sense of it together, asking every question under the sun to the guidance we were now both apprehending. Like those diagnosed with schizophrenia, we could hear voices. But the quality of the information, the inner knowing lighting us up and the strong physical sensations of expansion and bliss combined to assure us we were engaging in multidimensional communication. We were told that those who made the transition into this ability would have the responsibility of lighting the way for others. *The only true hierarchy is one of responsibility.* The price of wisdom is it must be shared. *The first shall be the last.*

One of the first messages we received was to use the phrase *I Am that I Am* to replace our negative mind chatter and quickly accelerate our growth. We did not understand yet that we were being advised to say over and over that we were one with God.

The I AM Presence is the ultimate power humanity has and *is* our individualised Godself. The name I AM was introduced into Judeo-Christian culture through the Old Testament Torah. In Exodus 3:14, Moses came across a burning bush and experienced it as a manifestation of God. Moses asked for this Presence's name. Translated from Hebrew, it said, '*I will be who I will be.*[145] *Tell the children of Israel I Am has sent me to you.*'[146] There is no present tense of the verb 'to be' in the Hebrew language. In the present tense, the phrase is *I Am that I Am* and this translation is what has come down the ages as the most powerful name of the ultimate Source, the closest we can come to having a name for this evolutionary ground of being and consciousness from which life emerges.

The beauty of the phrase I Am that I Am is that it is an ellipse of eternity: ongoing, infinite, non-temporal, self-existent, omniscient and omnipresent, rolling back onto itself forever. We issue from this as fractals from a hologram, the whole contained in its entirety in all its parts. This name is also an alchemist's crucible of infinite power – the seed-sound of manifestation, 'God in action.' It is the sound that represents All That Is, the infinite Prime Source and for each of us personally, the unbound Self. We are the Presence in action – and much as a wave is never parted from the sea, we are united with our origin, the infinite reality, enabling us to open the treasure box of its limitless abundance, the feast for the prodigal children who have returned home to union.

The great 20th-century Indian sage Ramana Maharshi, in discussion with Paramahansa Yogananda, proclaimed, 'I AM is the name of God. Of all the definitions of God, none is so well put as the biblical statement "I Am that I Am" in Exodus chapter three.'[147]

Another great Indian sage from the 20th century, Sri Nisargadatta Maharaj, also felt that the words *I AM* described the Supreme Reality that Hindus call Parabrahman, pure conscious awareness prior to thought or concept, memory, or associations, or the addition of identity.[148] Nisargadatta Maharaj's book *I Am That* is a revered spiritual classic contemplating the nature of I AM. He had achieved the ability to dwell in complete awareness beyond identity. 'I do not see the world as separate from me, so there is nothing to desire or fear.' He experienced himself as the All.

Ramana Maharshi and Nisargadatta Maharaj both taught that the words of enquiry, 'Who am I?' would finally lead to the ground of being beyond all concept, the I AM Presence, the ultimate expression of godhead. By continually asking ourselves the question, 'Who am I?' we come to realize we are not this body, this job, this partner, this culture, this gender, this race or these thoughts and emotions, *we are that which is observing them.* We are having an experience in this particular biochemical garment. Beyond all definitions of myself, I Am. When we find our seat in this consciousness, life flows very differently.

Sri Nisargadatta writes: 'Before all beginnings, after all ending – I am. All has its being in me, in the "I am," that shines in every living being.'[149] He says, 'Of course you are the Supreme Reality! But what of it? Every grain of sand is God. To know it is important, but it is only the beginning.'[150] And, 'Before the mind – I Am. I Am is not a thought in the mind; the mind happens to me. I do not happen to the mind. And since time and space are in the mind, I am beyond time and space. Eternal and omnipresent.'[151]

I Am That can only refer, in its ultimate purest sense, to that which does not pass with time, that which is non-dual, pure being, pure consciousness. We are moving our attention from what is temporal to what is eternal. And we will find, with our attention upon the eternal, that we are beings of bliss, awareness and consciousness in an infinite field of consciousness and infinite potentialities and that which is eternal permeates and informs all things.

Nisargadatta's own Guru and predecessor from his lineage, Siddharameshwar Maharaj, felt there were two roads to enlightenment: *The Ant's Way*, which involved lifetimes of meditation, or *The Bird's Way*, instant enlightenment catalysed by dispelling all thought with *I Am*. Nisargadatta was a bird, achieving ultimate liberation into unbound consciousness. Simply by asserting 'I Am' continually and with total focus and concentration, contemplating what is it behind the thoughts, he became able to dwell continually in non-duality with no perception of the illusion of separation, in permanent awareness of his eternal and perfect nature.[152]

The relevance of the powerful words *I Am* for those seeking enlightenment, whether by the ant's way or the bird's way, is immense. These words of universal power in action are how we accelerate swirls of electromagnetic energy into ordered activity, bringing the templates of Living Vision to fruition with passion, action and focused will. Sri Nisargadatta chose to move *beyond* desire and imagination. We who choose to continue to play in the world of form can still benefit from drinking from this cup of infinity.

The new dispensation of the Presence, or as it is called in the ascended master teaching, the Beloved Mighty I AM Presence or Magic Presence, occurred in the early 1930s. The ascended master Saint Germain, whose last earthly life was the European adventurer, scientist, linguist, composer, artist and master of the secret sciences of alchemy, Le Comte de Saint Germain, known as the 'wonderman of Europe' and 'the man who never died,' personally taught Guy Ballard, who had longed for contact with the masters of the higher octaves of frequency and had achieved the necessary harmony to tune the radio station to the right channel. Over a number of years, the information humanity needed to transition to a higher evolved iteration was imparted from Saint Germain. As explained previously, an ascended master is someone who has completed the Earth curriculum and who is has dedicated themselves to furthering our evolution and helping us ascend, willing prisoners of love. To ascend is to unite fully with first the Higher Mind, becoming like Christ or Buddha (or other fully enlightened masters) and finally uniting with the Presence, our eternal selves. By the way, this doesn't mean we need to don robes and sandals. This process is occurring in all demographics and walks of life.

Saint Germain works with other ascended masters and cosmic beings to free us from our hypnotic slumber. He laid down his momentum before the Universal I AM to have one more chance to relieve the intense shadows shrouding Earth caused by wars and continual destruction and conflict. This means he laid down his very life force and all his accomplishments to have another go at waking us up and delivering crucial information to help wake us to the realization we are limitless beings. The Presence is our force of life and as our union with it grows, so does our luminosity, our wisdom, love and power. When we have healed our emotions and become master of our thoughts and feelings; when we have begun to serve the highest good, our Presence seeks out these masters to tutor us. We may receive this in dreams, inner guidance or actual visual or audible conscious contact.

Your great Presence of Life, the individualized focus of God is a mighty Being of Fire that dwells with us in the realms of such exquisite perfection it would stagger the human intellect. [153]

– El Morya

Another great teacher of the I AM was Jesus. If we can shift our perspective of Jesus the Christ to a *wayshower* of the I AM with a teaching for all humanity regardless of faith, a very different picture emerges. His statements can be seen as I AM decrees and his merging with the I AM Presence is revealed to have been so complete his personal identity fell away. Much of his teachings have been distorted due to the misunderstanding of the I AM. He was so identified with his I AM Presence that his awareness was entirely focused there. He was not calling attention to the man, but the God in man, which he had fully become and which we will also. He had transitioned from Homo sapiens to Homo illuminatus and left a powerful blueprint for us to follow. *I Am the way, the truth and the life. No man comes unto the father except by me*, are revealed to be statements of the universal I AM, which *is* the way, the truth and the life and a call not to follow the externalized human creation, but the potential perfection of our individualized Godself, the way to unity with the universal I AM. *I Am the light of the world. Whoever follows me will never walk in darkness but have the light of life.* The light of the world is the Presence in action, the light of life expressed, the dissolver of shadows.

It is very heartening that contemporary inspirational authors and speakers including James Twyman, Neale Donald Walsch and the beloved late Dr Wayne Dyer have delved into the science of the I AM and shared their thoughts and insights with their readers and followers. Wayne Dyer referenced The I AM Discourses of the Saint Germain Press, a fountain of wisdom. Had he lived longer I believe this work would have taken center stage in his teachings.

When the words *I AM* are used, causal statements are transformed into powerful decrees. A causal statement is a statement that can cause a thing to happen. For every *cause*, there is an *effect*. For every action there is an equal and opposite reaction. This is a law of physics called Newton's Third Law. When our causes are rooted in the I AM, we are entering a very refined state of action and reaction. If our cause is motivated by love, our effect will be more love and this will manifest as the 'divine idea,' a template for manifestation that is an expression of our higher nature reflecting our own highest good and the highest good of all.

Whenever we use I AM at the beginning of a sentence, we are uttering a decree that increases this refined law of action and reaction, cause and effect. If we do it often and with feeling, we are commanding manifestation with our inner God authority and also building momentum. Therefore, we must use the words of power with caution! The universal force obeys the command of the sacred words and draws forth the substance of life to form your desire for you, especially when fueled by strong healed emotion. We are responsible for the qualification of every electron that passes through us. Life is given to us in a neutral state. If we misuse life, we must cleanse it and requalify it or it will return to us in ever-increasing discordant scenarios demanding our focus and often causing us to suffer. This is called *karma*. All discord must ultimately be transmuted for us to achieve liberation. To be liberated is to be in the perfect flow and alignment in continual connection with the Presence, not buffeted by the forces of our own miscreation. There are ways to accelerate the cleansing of discordant substance through the right use of energy, specifically the Violet

Ray, which you will learn more of later in the chapter and in Part Two, The Creative Alchemy Method™.

Our lives change once we begin to understand the words I AM as the words of ultimate creativity and animating power by which we may manifest whatever we apply our passion and power to. We are literally calling in the power of the universe. As we respect them, we become less cavalier. We become masters of destiny – as long as we manifest in alignment with our own highest good. It's a learning curve!

So, it's kind of a divine paradox. The words of power 'I AM' can dissolve you into the light of your true essence, reuniting you with the eternal sea and melting all desire. The words of power, correctly applied, can also help you manifest your heart's desire, enabling you to play more abundantly within this glorious dream called life. They can call forth the qualities of love, wisdom, power, peace, abundance, prosperity, health, creativity, youth, supply and joy when used sincerely and consciously in I AM decrees that cause you to embody the frequency of the quality, such as 'I AM love,' 'I AM wisdom' and 'I AM power.' We can call forth the tools of our ultimate purpose and our highest destiny and we can align with the most beneficial probability and call it into manifestation.

To seek first your true essence is not only wise counsel, it is also an alchemical formula and the key to the storehouse of our abundance! *Seek first the kingdom within and all things will be given to you,* means the place of your true essence, your I AM Presence *is also the place of your supply.*

That in whom reside all beings and who resides in all beings,
who is the giver of grace to all, the Supreme Soul of the Universe,
the limitless being – I AM THAT.
– Amritabindu Upanishad [154]

On a personal level, the effects of the spontaneous Bird's Way to enlightenment play ever more strongly on my consciousness, enveloping me in periods of quiet bliss or 'disappearing me,' like a wave that has become immersed in the sea. I emerge refreshed, nourished, aligned and ready to roll. Being immersed in the teachings of the I AM for 25 years at the time of writing this book (since the mid-1990s), I am vigilant with my words and thoughts and feelings. I have manifested a paradise very different from my early life which, colored by chronic violence and abuse, gave me great challenges to overcome, though there were diamonds in the mire. I fall off the horse from time to time and then I employ the techniques you will be learning in Part Two to work through what life has indicated is needing attention within me.

A Practical Understanding of the I AM Presence

You are a multidimensional being, converting cosmic energy into biological energy via your etheric subtle energy body into your energy centers to fuel your physical body, the mental body and the emotional body, which comprise your vehicle for your Earth journey. All the life force for every action, word, thought or feeling comes from the source of your life and the only force that can act in your life, your I AM Presence. Your Presence exists in a finer dimension or octave that is beyond duality where the frequency of perfection can be held.

Your Presence is a pure intelligent light, pure sentience and has all the qualities of the universal omniscient, omnipresent Presence from which life emerges. It is called your inner Godself because it is infinite, eternal and limitless. As you begin to merge with your Presence, you become aware of your unique tune and gradually become in perfect accord with the greater symphony of life.

Long before this final merging, the Presence can be consciously connected to just by intending and taking the time to listen through meditation and contemplation, gradually deepening and enhancing your life. We are all unconsciously connected to the Presence, we cannot fail to be, it is our very life force. The I AM Presence provides the electrical voltage that animates you and is the source of your authentic power. It is also the voltage that animates your manifestation.

To feel its love for you is a transformative experience that signals the end of seeking outside yourself for completion. At that point, you begin to draw autonomous people to you because you are evolving into one yourself. Their energy matches the energy you are emitting. Co-dependency, the state where we believe another is responsible for our happiness or unhappiness, ends when we find our own autonomy and sovereignty.

Love, a career, parenthood, friendship and whatever other activity we would typically pour our emotional body into, becomes more of a balanced giving and receiving than a search for something that can fill the 'holes' inside us, because our cup is full. You can learn to move with ease and grace between blended consciousness with the eternal sea and individuated consciousness, the local perspective of a wave upon it. Or, by mastering divine paradox, you may hold the perspective of both simultaneously as you harness the powers of the universe.

As the vehicles of your bodies become purified through the release of suppressed emotion, purification and self-care, you automatically begin to unite with the I AM Presence and ascend into it – or more accurately, it slowly subsumes your earthly bodies. You become less dense, shedding suppressed sticky emotions and outdated repetitive memories. Your awareness expands and you see more and with greater awareness. Patterns are revealed to you. What was previously veiled simply by being of a higher frequency (think of how the blades of a fan disappear when rotating swiftly) becomes clear and apparent because the oscillation (expansion) and vibration (contraction) of your energy field speeds up and becomes equal to it. These are the moments of transcendence when we can see and hear beyond the veil, and for some this becomes an ongoing ability. When you unite with this supreme Source of your being and begin to experience love of a very different scale, moments of startling clarity and the ecstatic feeling of pure life force pouring through you begin to occur. This increases inspiration, vitality, awareness and healing and also reverses or slows aging.

If you look at the photographs of people who have unified with their inner Godself, you will notice light emanating from their eyes. Check out Paramahansa Yogananda, for example. His eyes look like lighthouses! Or the beautiful Indian saint Anandamayi Ma. Once in a while I have seen this in myself and, in transcendent states, I have seen this in everyone around me. How is this possible? Light is the fabric of the universe. We are just peeling off the hypnotized layer of consensus reality that has taught us we are 'less' when in fact we are God-beings of infinite majesty in a sea of God, permeating everything.

The mystic and the physicist arrive at the same conclusion; one starting from the inner realm, the other from the outer world. The harmony between their views confirms the ancient Indian wisdom that Brahman, the ultimate reality without, is identical to Ātman, the reality within. [155]

– Fritjof Capra, physicist

Once you are sufficiently cleared and healed, you will begin to draw from the infinite power of your Presence in an increasingly conscious way. Your Presence is one with the field of infinite potential, when you have clear access to its power, the necessary tools and the unleashed power of healed emotions, a very big internal shift occurs. Life is no longer happening to you. You are happening to it.

Here's a story of when my body and my I AM Presence got together to shift something with no warning or mental preparation, although it was preceded by a humbly stated request. I had been working with the Presence for about 11 years. I have often found that 'newbies' sometimes skyrocket though, perhaps because they have less baggage and less to 'unlearn.' The timing is always unique to the needs of the individual.

My Bones Moved Like Putty

I was taking a break in Bali, one of most gentle and loveliest places on Earth. In Balinese culture, blessings are said for life all throughout the day and the veil between dimensions is thin. I was at the Buddhist teacher Burgs's beautiful retreat on the non-touristy eastern side of Bali. My husband had died, which was devastating, and my daughter had now grown up and left home. It felt good to be taking this break and I think it made the difference for me in making it through this intense period of my life, even though my responsibilities were piling up back home in London at the theater and pub I was now running. I tried to trust that all was well and gradually shifted into harmony with the loving Balinese energy. My environs were peaceful and empty.

One evening, as I was standing on the balcony of my little apartment overlooking the sea and the night sky, I observed that my jaw was snapping a little. I had always had temporomandibular joint disorder (TMJ), due to a forceps birth and so my jaw would build up tension and then crack. Snapping jaws are not uncommon in our stressed-out culture. But on this occasion, an odd thing happened. In a casual way, I asked my I AM Presence to fix my jaw. No ritual, just a request. I felt in a very relaxed state listening to the waves ebbing and flowing below and as I stared into the veils of luminous pulsing stars it seemed that the North Star was staring back at me.

Suddenly and in the most astonishing way my bones of their own accord began to adjust themselves. My jaw slowly moved forward without my effort and beyond any manipulation I alone could have achieved and with no direction from me. My tongue pressed on the roof of my mouth with the force of an elephant, spreading my skull. I couldn't believe what was happening. My bones were moving radically. This continued for an hour.

Being the irreverent person that I am, I saw dinner was being served down below and asked whatever force was moving my skull like it was putty if I could go down to eat. I felt the inner guidance give permission to go down with the caveat to avoid talking too much. Leaving the apartment, I passed the hall mirror and watched with fascination as the bones in my face continued to shift of their own accord without my participation. This was really happening!

During dinner, the movement stopped for the most part, thankfully! But when I returned to my room, it began again and went on until I finally fell asleep. This process carried on after I returned to London and for several months, though less and less. It often happened when I was in darkness and it could be suddenly triggered if the lights went out. The bones in my skull would begin to move.

Today the snap is gone. My chin is more forward, and my face is wider.

One of my takeaways from this experience is that the more we surrender authority to our I AM Presence, the more the Presence can act in our life. If we allow the I AM Presence to be the master of our mind, emotions, body and subtle energy field, it can reach beyond the range of our normally limited perception to help us.

We have four earthly bodies: mental, which corresponds to the air element; emotional, which corresponds to the water element; physical, which corresponds to the earth element; and subtle energy, which corresponds to the fire element. We have three cosmic bodies: our I AM Presence, which is anchored in our heart as the Threefold Flame; the Higher Mind, the bridge between the Presence and our earthly vehicle; and the Causal Body, the storehouse of our abundance and the energy field or aura of the Presence.

The Higher Mind, Higher Self or Holy Christ Self

In the wisdom teachings of the I AM Presence, the Higher Mind is referred to as the Holy Christ Self. As mentioned in the introduction, Christ is from Latin *Christus* or Greek *Khristós*, to be anointed – awakened by higher consciousness into your sovereignty. We are *all* fledgling Christs. This is a nonreligious term that exists outside of Christianity, although Jesus was a 'Christed being' who endeavored to teach us of the I AM Presence and awaken us to our true nature and was therefore named Jesus the Christ. I use the term Higher Mind to ensure this powerful knowledge is accessible to everyone, as it is intended to be. Krishna also means Christ. The Bhagavad Gita, a discourse given by Krishna to Arjuna, is an amazing record of timeless wisdom. We could also say Buddha Self or any other enlightened master who had achieved oneness with the Absolute. This aspect of ourselves is also sometimes called the Higher Self.

Another term for Higher Mind is *supermind*, coined by the great sage Sri Aurobindo, based on teachings from the Vedas. He stated that the supermind is Truth-Consciousness of the Divine Nature, an indivisible, non-dual state free of ignorance and full of light and knowledge. He believed supermind is the bridge between Satchidānanda and our four earthly bodies and it is only through the supra-mental mind that mind, life and body may be transformed. Satchidānanda means *existence, consciousness, bliss* and describes the subjective experience of the ultimate reality in Hinduism called Brahman – the supreme God force present in all things. This is the same illuminated wisdom from an Eastern perspective and semantics.

The Higher Mind is a stepped-down replica of your glorious I AM Presence, the cohesive power of divine love, wisdom and power that hold your mental, emotional, physical and subtle energy bodies together. When it is called into action, it will enlighten, purify and guide us. The I AM Presence only perceives that which is eternal and non-dual. The Higher Mind is the mediator between the perfection of our individualized Godself and the dance of polarity playing out in our dream of life. Unlike the Presence which sees only perfection, the Higher Mind can negotiate between the finer dimensions and our playground in the space–time continuum. See this part of you as your greatest friend. When we take the step of selflessness to co-create the shift of the ages and help transition humanity into its next evolution, we automatically receive the activating blessings of this aspect of ourselves and it steps forth into our lives When we call for the expansion of this powerful aspect of our being, we also begin to desire to call forth its expansion in all humanity. It increases its impelling power and releases the design for each one's perfected expression.[156]

Your earthly vehicle is a further stepped-down replica. If we remember that we are frequency beings, we can view these aspects of ourselves as existing in higher octaves that overlap and envelop the one most of us readily perceive. There is no separation.

Who am I? I AM that I AM.

When you connect directly into the sacred fire of your I AM Presence, especially if you have a lot of unhealed trauma, you are introducing a massive electrical volt into a blocked conduit. Calling for this friend and intermediary, the Higher Mind, steps down the massive power of your Presence until you are ready to contain it. If only I had known this when I began this journey.

The Power of the Field in Reverse

I had been in California immersed in the study and application of the mystery of the I AM. I was working with others, which really helps by increasing the power in the shared field exponentially, and many really remarkable things were happening that defied rational explanation. Our days were filled with miracles.

I felt my own electromagnetic field growing vastly during this time. I could reach out, sense and perceive in a very heightened way, very differently from before. To be in community in this way was a treasure and also really fun.

I returned to London in gray, cold winter after this extremely idyllic time. I perceived elements of my London life to be dense and at great odds with my recent experience. I was struggling. I had to go back into hiding or so I felt. I had barely begun to heal at that time. I felt my shadow emerge and begin to flood my now football field-sized auric field. The shadow self is the conglomeration of what is suppressed and hidden. It is often where treasures are but its sudden emergence can also derail you. What happened was extraordinary.

Just as seemingly miraculous events had been happening previously with the positive decrees I'd uttered and the state of being I had been holding for some months, I now created immediate manifestations that reflected my new negative state of mind, empowered by the strength of the field I had built up during my intense immersion in the I AM practice. I wasn't consciously decreeing negatively but my whole being was saying, I AM unhappy. As mentioned, the field is neutral and we can clothe it in harmony or discord.

People began to come up to me as I walked down the street and push me for no apparent reason. One time I went into a shop and the women in the shop surrounded me in an accusatory way, pointing at me and saying horrible things. Strangers! It was bizarre. It was a harsh but good lesson. My inner gloom was completely affecting my radiant field and, what's more, clearly bringing out the worst in others.

The Threefold Flame

Within the subtle energy field of your heart resides the Threefold Flame of love, wisdom and power. It is the divine spark where your Presence is anchored. This chamber is called the 'secret high place of the heart' in the Upanishad and the 'light of the spirit.' The Presence continually feeds primal life essence to you through this flame.

As we gain power this Threefold Flame grows, one day bursting forth to anoint you the 'light of the world.' [157] You and your Higher Mind have become one. When you picked up this book, it is not very likely you were signing up for emerging as a self-realized master. However, as you heal your emotions and align yourself step by step with your true destiny and higher purpose, this is an inevitable consequence of the path of the Great Work. *This is Creative Alchemy.* What does this look like in our modern world? Teachers and dancers and accountants and CEOs and athletes and writers and filmmakers and bakers and plumbers and scientists and hairdressers and artists and gardeners and inventors and stay-at-home mothers. Whatever we choose, as long as it is in alignment with our purpose and destiny. But maybe if you more closely, there's that light in the eyes! The moment you intend to hitch your wagon (your earthly vehicle) to a star (your Presence), the wheels of motion begin to turn and you become a wayshower too.

What does this have to do with Creative Alchemy? The likelihood is that you picked this book up because you wanted to heal your emotions and/or you wanted to get better at manifesting the life you want. We're still on track. What you are learning here will be applied as super-tools in Part Two.

When we get started you will see it's as easy as building a brick house. Just one brick at a time. The rewards are great. Before long your new structure is built. And then one day you'll take down that house brick by brick, when you and your Presence are more fully united and everything just flows. The structure is then inherent and needs no external props.

The Causal Body

A crucial aspect of your multidimensional being that will assist you to experience your ideal destiny is your Causal Body.

Your Causal Body is a rainbow force field of seven concentric photonic energy bands surrounding your I AM Presence. It is a battery of energy that provides the momentum for your manifestation. In its various bands, all the positively qualified electrons and fundamental particles from your experiences in all timelines and dimensions are stored. You manifest from this storehouse when you achieve *harmony*. The original Aramaic translation of the beatitude, 'The meek shall inherit the earth' is 'Those in harmony shall inherit the earth.' The colors of the bands represent different qualities or spheres of consciousness and experience. As you develop these bands become clearer and brighter.

As you clear blocked energy from your mental, physical, emotional and subtle energy fields, you will come into greater and greater harmony and the Causal Body will automatically release an abundant store of gifts and talents to you with which you may enhance your life and the lives of others, as well as provide the supply of abundance you need. As you connect more deeply with the Presence, your Causal Body will release more and more of its substance to be moulded into form.

The size of the Causal Body is determined by the amount of energy you draw from the universal storehouse which is determined by how we spend our abundance. How are you expending your energy? Are you developing your gifts, powers and talents? Do you relate to others with compassion and tolerance? Are you aligned with the highest good for all and your own highest good? Are you keeping in balance and inner peace? Do you send healing light where there is strife? To you tithe? Are you generous when possible? If so, you are building your Causal Body.

The keys to our release of abundance and the fulfilment of our dreams are states of harmony and peace. You can discern that you are in harmony when your emotions are clear and your perceptions are uncolored by the past, when the movie of the outer world no longer rocks you and your inner truth is stronger than what appears to playing out in the dream of life. When you are in harmony the frequency emissions of your energy body will also be. Your cells will resonate the song of your highest good sending a signal to attract and manifest your deepest, most authentic and rewarding expression. Inspiration for specific actions will come to you and synchronicities will begin to flow. The rewards are great. Getting you to this state of equilibrium is the goal of Creative Alchemy. When you are in harmony your Causal Body can release the codes of your true purpose and your storehouse of abundance will begin to flow.

The seven concentric spheres also relate to the Seven Spheres of embodiment, each imbued with different qualities. The Sphere you have spent the most time developing in will also be the largest band of your Causal Body. We often relate to a combination of a few. Those of us who seem very single-minded and focused on a direction will more likely relate to one specifically. Sometimes our favorite color can also be an indicator.

Remember color is wave frequency, a consequence of energy produced by vibration and electronic transitions, visible light/energy. All energy holds information.

A Very Brief Overview of the Seven Rays or Spheres of Learning

First Ray: This is the realm of the Will. Color, blue. When the Will of the four earthly bodies is surrendered to the Will of the cosmic bodies, divine ideas are born. When those with an affinity with the First Ray are underdeveloped, aggression and domination can manifest. Service: rulers, leaders, protectors, company founders.

Second Ray: The realm of wisdom and illumination. Color, yellow. Where ideas (imagination) are perceived and moulded into patterns that can manifest as workable forms (Living Vision). When this ray is underdeveloped, it can lead to intellectual arrogance, spiritual pride and an excessive rigidity and fixation on worldly knowledge which presents as an inability to access higher inspiration. Service: teachers, educators.

Third Ray: Immortal Love. Color, pink. Compassion, tolerance, gratitude. Immortal love is the cohesive magnetic quality that sustains the form of conscious manifestation. When it is underdeveloped, it leads to lack of love and compassion for other beings. Service: peacemakers, arbitrators, diplomats, artists.

Fourth Ray: Purity. Color, white. Purity is a substance and a feeling. In holding the 'Immaculate Concept' for others and ourselves, we imbue their individual unique plan with this special substance that maintains the integrity of the template.

Fifth Ray: Consecration and concentration. Seeking higher truths. Color, emerald green. When it is underdeveloped, it leads to doubt, discouragement and lack of discipline and focus. Service: doctors, engineers, nurses, scientists, inventors, artists.

Sixth Ray: Devotion. Color, ruby tinged with gold. When underdeveloped, it can lead to religious fanaticism and zealous tendencies. Service: spiritual leaders, prophets, thought leaders. Our last 2000-year era was a Sixth Ray era and has been characterized by a great spiritual awakening, but also the flip side of fanaticism and over-zealousness.

Seventh Ray: Alchemy! Color, violet. This is the ray of transformation and the ray of our incoming era. Underdeveloped people on this ray can fall into snobbery and lack of true refinement of spirit. Service: diplomats, ministers, sages, healers, quantum theorists, artists, shamans, masters of energy.[158]

The Column of Light

The Column of Light is the tube of protection you can call down from your Presence every morning and evening. As it builds up strength, it shields you from random thoughts and the energetic onslaught of modern life and even the non-beneficial emotions of others. It holds your vibrational frequency at the highest level possible and these faster oscillations of energy repel the slower, denser frequencies, protecting you from discord. It should be visualized as nine feet in diameter and is also called the *canopy of sacred fire.*

The Alchemist's Ray – The Violet Flame

Through and around your body you can call forth the ceaseless activity of the Violet Flame, an activity of the Seventh Sphere or Ray. Calling the Violet Flame to cleanse your four earthly bodies will greatly accelerate your progress as you strive to improve the course of your destiny. You could call it energy hygiene, just as you cleanse your physical body with soap and water and walks in Nature, your emotions with Creative Alchemy techniques you will soon learn in Part Two, your mental body with meditation, music and positive literature, the most powerful technique for cleansing your energy field is the Violet Flame.

This is a highly calibrated, alchemical activity of light which transmutes discord into a neutral state which can then be qualified with love, wisdom, power, victory, abundance and so on. Violet is the highest frequency visible color at 750 THz. THz (terahertz) is a measurement of electromagnetic wave frequency equal to one trillion Hz per unit. The Violet Flame we are calling in is beyond the spectral range and is an even higher frequency. You may find that with determined use of the cleansing power of the Violet Flame you begin to see flashes of violet light. It is *the fire that does not burn but transforms all it touches into itself.* As we use this energy to cleanse our energy field, we lessen the impact of accumulated debris on our physical, emotional and mental bodies as well.

Colored light has been used to heal by ancient cultures including Egypt, China, India and Greece and in indigenous cultures. In 19th-century Europe, Dr Neils Ryberg Finsen won the Nobel Prize for his pioneering use of ultraviolet light to heal tubercular skin lesions, and many other practitioners followed his success using rays of colored light as a powerful healing tool.

The Violet Flame protocol is twofold. First, we call the Law of Forgiveness for our self and others for the misuse of the life force, whether accidentally or consciously. The Law of Forgiveness, if applied sincerely, heals separation and restores unity. Secondly, you call the Violet Flame to bathe you, lovingly commanding it to dissolve *cause, core, record, effect and memory* of the discord you or any other has created in any timeline or dimension in action, thought, word or feeling. It is wise to call a guardian of the Ray to help you sustain its activity. We are assisted by an infinitude of beings from the cosmic, elemental (the spirits or energetic informational essences of earth, air, water, mineral, animals and plants), ancestral and ascended realms. There are many who are dedicated to ensuring our liberation from the hypnosis of limitation, the spell we currently believe to be reality. Saint Germain and the Archangel Zadkiel are two you can call for support in anchoring the cleansing activity of the Violet Flame.

The archangels are vastly intelligent beings who hold key qualities. They are great masters of the sacred geometry that precedes manifestation and work with the builders of form to assist in the birthing of constellations and galaxies, planets and biospheres, flora and fauna, and the great universal cause of manifesting consciousness into myriad forms. It is humbling they will answer our call and assist us. Humanity stands midway between the angelic realm and the mind of nature, the elemental realm. When we can work in harmony with both these forces we become greatly empowered.

Metaphysics and physics continually grow closer. True science is the quest to understand the laws of our mathematical, musical universe of which these beings are a part. Everything is intelligent energy, waveforms interacting. When we can accept this energy as consciousness in different forms and call upon the help that is there, the treasure chest of the universe begins to open. We can return this help by sending the Violet Fire to people and places in need, completing the Law of the Circle.

The Violet Flame practice greatly reduces the accumulated pressure of discord you have built up, which can appear in the etheric as thick tar, gradually reducing to smoke and finally fading away with repeated practice. Ridding yourself gradually of this accumulation helps with changing the patterns of your energy field and laying down of new neural pathways. It is a cleansing which eases karma, the return effects of your causes and is therefore infinitely merciful.

Electrophotonic imaging cameras are able to photograph energy and can record the color spectrum of energy not visible to the human eye including the seven main energy centers of the body. It is currently being used by thousands of professionals to aid in the diagnosis of the health of individuals by analysis of the human energy field, clearly recording the various colors of our energy centers as well as size and placement. It won't be long before it is put to use recording these multidimensional aspects of ourselves. I created my own experiment with one of these cameras, calling in the deep pink ray of divine love, the violet fire and golden light. Though not a controlled study, the results were interesting. All three colors appeared in the energy field around my head. This photo is on the Creative Alchemy website, www.creativealchemy.vision

The understanding of the I AM Presence and the transmuting power of the alchemist's ray have long been only available to masters who dedicated their lives to apprehending the true inner teachings, often by isolating themselves in ashrams or caves. Alchemists and healers throughout time have used these powerful tools. It is the dispensation of this age that these teachings have been made available.

We are prismatic beings of light and each ray of light and shade of color holds qualities that can affect reality when focused. Every electron and particle you draw from life will be qualified for constructive or destructive use. Those qualified for beneficial use come back to your bank with interest. Those unqualified or misused will either render karma, the effect of your cause *or* by this special dispensation to assist humanity at the time of this great shift of the ages, be cleared and restored to their original perfection by the use of the Violet Flame. Remember every particle is on its own journey to mastery. Walk through your life with consciousness as to how you treat everything and everyone.

A Visual Model of the I AM Presence

In an effort to create a linear model of what is nonlinear, multidimensional and transcendent, visualize your I AM Presence as being anchored at your center – within your heart – like a gazillion-megawatt, glowing sun of energy flooding out in all directions all around you. Every small and large system in nature organizes itself around a central hub of power, whether that is the sun in our solar system or the nuclei at the center of our atoms. This model leads us to look deep within to have our needs met and expresses the natural reality of both the macrocosm and the microcosm.

For millennia we have been taught to look outside ourselves for answers, blessings, wisdom, forgiveness and supply. Even looking to the Universe implies the answer is outside you. Your teacher is within. Your bestower of blessings is within. Your fountain of wisdom is within. Your power to forgive, the power that makes reparation and heals separation, is within. Your supply of all good things is within. All the dimensions of your being are within. When you become completely united with the love, wisdom and power within you, you also become united with infinite Source. You become a sun. This is our goal, to be completely unified with the Presence – to embody it.

Each Presence has a unique pattern, which is as unique as a thumbprint, and a unique musical keynote, the song of its reason for being and its service to the universe, within which is held your mission, your purpose, your special gifts and talents.

When your personal endeavor to master your thoughts and your feelings has been sufficient, your Presence calls forth the higher guidance you need to evolve and you will meet these helpers who are dedicated to your liberation with love and compassion either by sensing, seeing or hearing. There is never a time when we are not guided, we are just unaware. When the lion's share of humanity master this, we will solve our problems quickly. This activity is crucial for our future. As Einstein stated, *no problem can be solved on the level it was created*. We must evolve.

Many cultures have never lost the ability to see beyond the veil of third-dimensional density, especially indigenous cultures that cultivate authenticity and intuition and do not find communication with these guides and helpers exceptional. They look upon us as if we are blind and to some degree we have become so. Please go to www.creativealchemy.vision to see an image of the *I AM Presence*.

Delving Deeper into the Science of Decree

You shall also decree a thing, and it shall be established to you. Light shall shine on your ways.[159] What exactly does this mean? And what exactly is a decree or the act of decreeing?

A decree is more powerful than an affirmation, although both offer a blueprint for manifestation just as an architect's drawing envisions a house before it is built. The reason a decree is infinitely more powerful than an affirmation is because it uses the words of power, *I AM,* the 'God-in-action' words which draw forth the immense power and life substance of your Presence thereby animating your thoughts and imbuing them with powerful life force. This sustains its energy and keeps it active long after we have forgotten the call we have made.[160] When a fiat or decree is made in the name of the I AM it lives eternally.

The clarity of the visualization creating the geometric pattern around which the life force energy can coalesce forms the greater part of the power of the decree.[161] The charge of positive confidence and feeling empowers it and is the battery or fuel. When we decree from the core of our being with the correct protocol, the decree will manifest.

It's important to note that *wherever* our attention is we are guiding the substance of life to manifest. Be careful where you place your attention! If your attention is on strife, then use the Violet Fire to send healing waves lest you signal the manifestation of strife in your life.

Harmony + Clarity of Vision + Purified Emotional Fuel +
Power of the Presence + Action = Manifestation

The time its journey takes from your heart to full realization is dependent on clarity, conviction and universal timing. Conviction is crucial in technology of mind. This is what is meant by the biblical passage: *For most certainly I tell you, if you have faith as a grain of mustard seed, you will tell this mountain, 'Move from here to there,' and it will move; and nothing will be impossible for you.* [162]

The decree is the rain watering the abundant garden of the dreams of your heart. With application, it will bloom and bear rich fruit.

Imagine we are swimming in a humming energy soup. We are! It permeates us and surrounds us. It is vastly intelligent and vibrating with infinite life, infinite potential. Everything we say and feel, especially with a strong emotional imprint, encodes this unified field. It instantly responds by forming architectures of geometry. Then the vast manifesting sentience of it says, 'Okay, thanks for your order today. Comin' right up!' When we decree with power and emotion, we clothe those geometric patterns with pure life substance for manifestation into form in the physical realm by magnetizing what is needed.

When we become truly aware of our thoughts on the conscious and unconscious level, we can learn to substitute negative thoughts with decrees and mantras, erasing repetitive patterns of thought. Becoming responsible for our thoughts, feelings and actions is crucial.

It's a good idea to use the Violet Flame once in a while to mop up all half-baked ideas, unmanifested imaginings good and bad that will be floating around in your etheric field. These may not make it to the physical realm but they are still the children of our psyches and they can affect the energy fields. If it's in your life, you created it. If it's in our world, we co-created it.

If you want to understand where your consciousness is, simply look at your present life and take stock of what has been out-pictured from your inner world into being. Let's revisit the questions in Chapter One.

How are your relationships?

How is your health?

Do you love what you do and where you live?

How's your inner-peace quotient?

What about your happiness quotient?

Are you in a position to make contributions to the greater whole? To pick up your perfect instrument and play your unique melody in the great orchestra of life?

Are you part of a meaningful, nourishing community where you can be your true self?

How are your finances?

Do you have meaningful, purposeful work?

Summing Up the Power of the Presence

Our Presence is the generator of our life force, our powerhouse, the animating factor of our life. When we call on the Presence using the science of decree and the activating words of power, the illuminating power of consciousness itself can act in our world. It has three main parts, the Presence, the Higher Self and the Causal Body. Within the heart, where the Presence is anchored and flooding us with life force is the Threefold Flame of love, wisdom and power. Our Presence projected this sacred spark to manifest our four earthly bodies. You're going to learn in Part Two that this process is one of the great secrets of true manifestation.

- Use the words I AM to form a decree to invoke (call in) qualities you want to experience in your life. Seal your decree with the sustaining activity of peace.

- Call for the highest beings possible to be your guides and to make themselves known to you.

- Call to your Causal Body to charge your being and your world with all its good qualities and energy for your use NOW to enhance your life and life in general. Stay in harmony!

- Call to your Higher Mind and ask it to charge your four earthly bodies with its consciousness. Do this morning and night.

- Call upon the powerful alchemical ray of transmutation, the Violet Flame, to transmute all cause, core, record, effect and memory of patterns that disturb your life. Call upon it to cleanse every electron ever misused by you in all timelines and dimensions. Help others by calling the Violet Flame to flame through all life on Earth to relieve the pressure of discordance so more can experience happiness, harmony and well-being. Always first call on the Law of Forgiveness to heal separation, restore unity and acknowledge our awareness that we have harmed life, either accidentally or on purpose and call on a guardian of the Ray to help sustain the activity.

- Release all who have ever harmed you from the consequences. Everything is a mirror and a learning experience.

- Call down your Pillar of Light and protection every 12–24 hours. Flood it with the fire of the Violet Flame so all energies coming towards you are cleansed as well as any and all energy you send out.

- Give thanks. Create a habit of gratitude. Thank everything in your life, even inanimate objects. Everything is made of electrons and other particles of life which at their core are waves of intelligent energy.

- What's your favorite color? Which of the seven Rays do you feel an affinity for? It's often one or two. See what you can do to develop more of the qualities of your Ray.

- The decree *I AM harmony* is very powerful and helps to organize universal forces into beneficial patterns. It can be used as a mantra and can help change negative thought patterns and inner mind chatter.

PART TWO

THE PRACTICAL APPLICATION OF CREATIVE ALCHEMY

THE CREATIVE ALCHEMY METHOD™

TECHNOLOGY OF EMOTION – MASTERY OF LIVING
VISION – CONNECTION TO THE MAGIC PRESENCE

PUTTING IT ALL TOGETHER:
TEMPLATES FOR MANIFESTATION

The Practical Application of Creative Alchemy is presented in four phases.

Phase One: The Technology Of Emotion

You will learn how to free suppressed emotion and unleash your power. Using simple but very powerful alchemical creative techniques, you will be taken on a journey of emotional discovery. You will be given techniques for freeing and healing emotion and restoring suppressed emotion to *energy in motion*. This cleansed energy literally becomes superpower to fuel your life. Releasing suppressed emotion increases your vitality, your natural self-healing abilities and greatly empowers manifestation. This work can be powerfully cathartic, but there is no right or wrong result. Each of us is unique.

You will gradually become more available and *present* for life, love, friendship and opportunity. You will feel increased vitality and clarity. You will be triggered by external events less and less and then finally not at all.

Phase Two: Living Vision

When the imagination is elevated and released into the Higher Mind where unity consciousness prevails and empowered by purified emotion and clarified will, we begin to operate as masters. You will be given many special techniques for building your ability to imagine and for transforming your imagination into Living Vision.

Phase Three: Connection To The Magic Presence

The Presence is the source of your life and the *only* consciousness that can act in your life. This ancient science is the most powerful and crucial piece of the puzzle although all steps must be taken. You will be given techniques to connect with your I AM Presence and the Threefold Flame anchored in your heart which holds your superpowers of immortal magnetic love, illumined wisdom and true power. This will begin your journey to sovereignty and autonomy and the release of your own inherent supply of abundance. You will develop the ability to receive your own highest guidance or enhance it. You will begin the transition from Homo sapiens to Homo illuminatus.

The electrical life force of the Presence combined with the science of decree are the final ingredients for a method of manifestation once considered too powerful for the masses, remaining hidden and available only to those who had proved themselves worthy of it, who through enormous discipline and perseverance, underwent arduous initiations gaining enough wisdom directly from the inner planes to be deemed worthy of illumined master teachers. It is time now for us to step up and become ambassadors for a new world and learn to use these great gifts for our own freedom and the freedom of all.

Phase Four: Putting It All Together – Templates For Manifestation

In Phase Four you will learn how to put it all together and become a conscious creator, a creative alchemist, the master of your destiny using a proven process. These templates have been manifested through my own inner-plane work combined with years of study and training with master teachers and collating vast amounts of material to create an accessible system for modern lives. Templates are provided for the qualities of Joy; Health, Youth and Beauty; Prosperity and Abundance and for Relationships. You will begin to get the feel of it and start to tailor it for yourself with the guidance of your Presence and Higher Mind. You must give yourself time and be diligent.

After you have completed these four phases and learned to apply the Creative Alchemy techniques as needed, you will have a sound practice to carry through your life. You will get better and better at manifestation until one day it is completely fluid. You embody it. You are finally surrendered to and united with your Presence. As described by Swami Sri Yukteswar, *the state where we move freely in the world performing our outward duties with no loss of God-realization.* You will naturally turn your attention to co-creating and assisting the whole and experience the rewards of this. Through conscious co-creative collaboration, we will innovate solutions to restore our beloved Earth to the paradise she was intended to be and assist all humanity in evolving. You will fulfil the dreams of your own life.

You can anticipate becoming powerfully aligned through this process. The system in its totality will enable you to become congruent physically, mentally, emotionally, energetically and behaviorally and will ultimately align you with your Source, the Presence, on a continual basis. This is the part of you that holds the blueprint for your highest destiny, your gifts and talents, and is the aspect of you that is connected to *infinite wisdom, infinite power and infinite love.*

For full efficacy, Creative Alchemy should be approached step by step, beginning with Phase One. The healing of emotions is by far the most important step. If our emotions are released and healed, we enter a heightened state of harmony and everything else will come into alignment with ease and synchronicity. You can learn to manifest without healing your emotions, but it is unlikely your manifestation will sustain itself or that you will find the fulfilment you seek. Secret feelings of doubt, unworthiness and other emotions can create sabotage. If our emotions aren't cleared, we can also fall prey to manifesting out of alignment with our highest good which has its own consequences.

I have applied this practice to myself and others for a quarter of a century now and patterns have emerged. Life spirals up and down. It isn't linear. We make amazing progress and then we become tired or feel defeated or an event knocks us out of alignment. We may then spiral back to our default position where the lost children of our psyches are waiting to greet us. But the next time we enter the upward spiral we do so with greater strength, integration and understanding and each time more of these lost children are resolved and integrated. Our growth as people may plateau, but rarely do we find ourselves in the distraught or blocked states we entered our practice with initially. *We do not lose the momentum gained* even if we momentarily stall. Never lose heart.

My hope is that you will fashion an ongoing practice out of some or all of the steps. A practice is, as the word implies, an ongoing transformational technique for life. The rewards are great.

To manifest our true purpose, we need clear, clean emotions with nothing lurking buried – and we need to fan our positive emotions into a heightened state. The passion you flood your vision with is the vision's lifeblood. True passion for the purposes of Creative Alchemy is *focused inspiration*. Passion is the motivator ensuring you take all the action steps necessary and don't just create castles in the clouds.

I heard the old maestro of the guitar say: 'The duende is not in the throat: the duende surges up, inside from the soles of the feet.' Meaning, it's not a question of skill, but of style that's truly alive: meaning it's in the veins: meaning, it's of the most ancient culture of immediate creation.

– Federico García Lorca, *Theory and Play of the Duende*

Note: *Duende* means a quality of passion, spirit, inspiration and a heightened state of emotion, expression and authenticity.

The quote speaks of an understanding that emotion fuels creation and also that passion is even more important than skill. Skill can be learned. Do you want an amazing voice? Use Creative Alchemy techniques to bring your dream into sharp clear focus, ensure you have the requisite passion, take action as guided. This same formula works for whatever your dream is. It is never too late. Now, let's begin with Phase One of the process: clearing suppressed and blocked emotion.

CLEARING AND HEALING SUPPRESSED EMOTION AND UNLEASHING POWER

When inward tenderness finds the secret hurt,
pain itself will crack the rock and, ahhh!
Let the soul emerge!
– Jalal ad-Din Muhammad Rumi

The basis of the healing system we will use in the practical application of Creative Alchemy is built upon the understanding of suppression and expression we discussed in Part One. What is suppressed causes havoc. What is expressed creates insight and relief when it is released and makes life force available as personal power. Memory is stored in images. The alchemy is in the creative expression.

With guided intention, the lead of suppressed emotion may be transmuted to the gold of increased vitality and well-being. Powerful, heightened, positive emotions can now emerge that can heal our bodies and minds, fuel manifestation and raise our energy fields, heightening vitality and magnetism.

It's possible you will have a very strong emotional reaction at times as you access buried feeling, even a full-blown cathartic response. The best way to describe a cathartic release is as 'pain on the way out' – and it feels great. If we haven't experienced it, we are often afraid of it which is one reason why we suppress emotion. Many angry people are afraid if they let the angry feelings out they will commit mayhem, possibly harming themselves or someone else, perhaps someone they love. So, instead of catharsis, they hold back and let the anger leak out, scalding people around them. Or the anger is expressed as anxiety, frustration, annoyance, passive aggression or irritation. If deeply suppressed it becomes chronic depression.

Too many of us have been told that we need to 'keep a lid on it,' or words to this effect. Except putting a lid on emotions will turn you into a pressure cooker. Some people feel that if they release their tears, they'll create a flood. Or worse, they will expose their vulnerability and be rejected and abandoned. We need safe ways to release what's buried in us. It's crucial to do this work! The ultimate issue is that we cannot suppress selectively. If we are suppressing anger, fear or grief we will also be suppressing joy, peace, excitement and calm. We will feel deadened, frozen. Our aim is to feel alive, warm and vibrant. If you reach catharsis, congratulate yourself and keep going! It can be subtle or extreme, there is no right or wrong in it.

Our first exercise is a practice that helps us access the biography of our biology, our living library, and to embody our emotions in a visual way.

If you are feeling overwhelmed it is better to have someone with you. If you are having trouble coping, having harmful thoughts, have a history of emotional and mental difficulties, or receiving medical or psychological care, please check in with your health-care practitioner.

Exercise 1: Make a Visual Representation of Your Emotions

Let's dive in and grab an emotion through painting.

Supplies: Washable poster paint, a tarp if you are working over a carpet, some inexpensive brushes, some old clothes and thick paper able to take water. All of this can be purchased cheaply in the kids' section of a stationery or art shop.

Time: 5–10 minutes per painting. Usually two to three paintings per sitting. Take as long as you need. Give your self time!

Preparation: Seal the page. Simply say, 'I bless this page' and no energy will travel from this exercise. As energy does travel, sealing and blessing the page confirms that the images and narratives are for release and healing only. Energy responds to intention. Understanding this is a step towards becoming masters. Center yourself, take some deep breaths and become aware of your body.

How to do it

Step 1. Scan your body and sense any areas that feel different. You are scanning for a predominant emotion but it may show up as an irritation, a buzz or just draw your focus with a different quality or feeling from the rest of you. Take your time. When you have found an area, breathe into it. If you can't find an area at first, don't worry, as you repeat this exercise your ability to sense your body will grow. Your body and emotions want to cooperate. They have been trying to communicate. If you have breathed into your body and perceived an emotion, write it down on the corner of the paper. If you haven't been able to feel into your body just yet, then without thinking write down in the corner of your paper the first emotion that comes to mind. Trust that your body will give you one. Writing the emotion is another signal to your body to release.

Step 2. Without thinking or hesitation, choose your colors.

Step 3. Next, without thinking, staying with your body and the feeling, (or focusing in your body), paint a picture of that emotion. It's important not to think – just begin to paint spontaneously. You are painting a feeling. Keep going until you feel ready to stop. This image can be a picture of something, but it is almost always abstract. A mess of colors representing the emotion.

Step 4. Put that painting aside and begin another one, repeating the above process. You can also stay on the same paper. This time, drop the brush and paint with your hands. This connects you directly to your body and allows fuller immersion. Most often these patterns of repression begin in childhood, and using your hands as a young child would do can help to trigger release. Repeat one more time. Of course, if you want to keep going, please do!

Having facilitated this exercise countless times with individuals from six years old up into the eighties, I want to reassure you that whatever happens is perfect. I have seen paintings of despair, hopelessness, rage, joy, grief, fear, dread, anxiety, irritation, anger, joy, peace, calm, excitement and other emotions painted by people who have never put paint to paper in their lives. This exercise breaks the dam of suppressed emotions. It is often the second or third painting that starts to deliver a stronger emotion, so be patient with yourself. Frequently the first painting will be happiness, irritation or mild anxiety.

You may continue to paint on the same sheet of paper effectively transmuting the visual representation from one emotion to another. This can be very powerful. Despair may turn to hope, for example, or grief into rage. Don't impose any direction. If you are truly feeling joy or happiness or peace, by all means paint that.

Remember, no emotion is bad emotion. Stuck emotion gives us trouble.

Keep checking inside your body for a feeling – *emotion is always something we can feel in our bodies*. If you have been sincerely engaged in a healing journey and have cleared a lot of patterns and emotions or if you were blessed with a childhood that left you with a very happy and harmonious disposition, I still encourage you to do this exercise and the others in Phase One. You are giving your body and emotions loving attention and permission to be heard. This is invaluable.

This exercise is very good for relieving anxiety, our modern malaise, which is caused by suppression and lack of authentic expression. It can also help resolve depression and other forms of mental and emotional distress. Clearing our suppressed emotions creates relief and makes space for greater happiness and well-being. This process allows the energy to be released from your body and leaves you with a more complete open state of being. This energy becomes your superpower and is life-changing.

The psychoanalyst Wilhelm Reich called the effects of suppressed emotion 'body armor.' We create this armor to keep the emotional pain hidden and to shield ourselves from further pain. It is a literal hardening of the body that can proceed to illness and aging. Emotions are information-packed energy and when they are blocked other systems will also stagnate.[163] The ancient Chinese energy masters discovered blocked emotions are stored in our organs. Fear in our kidneys, anger in our liver, grief in our lungs and so on. Releasing blocked emotion is crucial for mind, body and emotional well-being.

When to use this

Continually until you feel clear and then when you feel out of alignment, or when a big trigger has occurred or a traumatic event that you need help coping with. Below are some examples. Remember, always trust the way your earthly bodies present a memory. If it goes against your current belief structure, just think of it as a metaphor.

Examples from Clients' Experiences

Rose was a lovely young girl confined to a wheelchair who also had difficulty with verbal communication. Her caregiver supported our session. It took some time for Rose to make her first painting. She felt very shy and intimidated. She finally made a huge angry painting with wild strokes of red and black. Tears of anger fell. The caregiver and I held our breath. Rose finished and closed her eyes. Gradually a huge smile spread across her face, which had relaxed considerably. She high-fived us. How much rage must she have been carrying with the challenges she faced? She now had a tool to express this huge emotion.

Janet, a 30-year-old woman, scanned her body and found a feeling of irritation in her chest. Breathing into it a memory surfaced unlocking tears as the trauma of having her creativity be suppressed by a teacher early in her life re-emerged, along with the pain of losing an important part of herself – her ability to express herself and her authentic creativity. Her teacher had criticized her artwork and told her to think of doing something else with her life. Young Janet took this to heart and shut down the creative side of her personality. She was encouraged to move back into the feeling in her body and experience it fully.

Her first painting was of pure grief. Her second painting was anger and frustration. Janet kept going with the process, discovering how releasing these emotions unraveled a ball of tightness she had carried in her stomach as long as she could recall, causing digestion problems and often forcing her to shut down with others. Over the next few months of just using the first two steps of Creative Alchemy diligently, Janet continued to expel buried emotion and befriend parts of herself she had abandoned. Her newly unleashed power has increased her vitality and radiance significantly. Her stomach issues have cleared entirely.

A group of teachers at a school first expressed themselves politely and on their second attempt to make a painting began to release strong buried emotions. Seeing what their colleagues painted and that they were not alone in their emotions helped the faculty bond in shared vulnerability and recognition.

A scientist accessed beautiful creative abilities and was later able to illustrate some of his concepts. However, this is absolutely not about making beauty. If you consciously try to make anything you will jump into your head and not be able to access your emotions. It's more about accessing power.

Countless individuals have found themselves back at the site of trauma releasing physical and sexual abuse and other trauma, including those who felt sure they had healed their wounds.

We can heal in one or more of the four earthly bodies while neglecting a powerful imprint in another. Through the power of creativity to engage all four bodies we can find these hidden traumas and release them.

Adults from all walks of life have been able to release buried emotion with this one simple exercise and are often amazed by what is suppressed and how easy it was to access it. Very rationally oriented people can feel resistance at first but are then surprised by how effective it is.

Through this exercise young children have released powerful emotions that their parents and teachers had no idea they were suppressing. It has become an important tool in expressing how they are feeling. Usually children and even adults don't understand the difference between a thought and an emotion. An emotion is always felt in the body and then perceived by the mind. Children have been helped with obsessive–compulsive disorder and adults and children have been helped with anger management and attention deficit disorder.

I recommend you give yourself the time and the space to engage in this exercise until you feel you are becoming very aware of your emotions and you have begun to release them. Signs will be greater physical comfort, reduced anxiety and much less mental stewing and repetitive compulsive thoughts.

Our thoughts and emotions are continually creating our life. There is never a moment when we are not creating as we are thinking and feeling even in our sleep states. When we get in touch with our inner landscape through authentic creativity and release suppressed emotions, then we can more easily align with the states of heightened positive emotion necessary to manifest our true purpose. Eventually we create our life in easeful flow as an inevitable consequence of our inner alignment. We follow the guidance we can now clearly perceive and take it step by step, living more in the present with less and less regret about the past or worry about the future.

Once you have begun to become physically aware of where various emotions are stored in your body and you are releasing them freely, feeling them deeply but letting them flow, you can save this exercise for when you need it. The calmer, clearer and more harmonious we become inside, the calmer and more harmonious our external reality also is.

Sometimes in painting an emotion a memory will come up. Sometimes the painting is of an emotion without any thought or memory attached. If a memory comes, that's great. Breathe into it. Bring your focus gently back to the feeling of the emotion and return to painting.

Note: If the patterns of your life have landed you in an abusive relationship, you may need to take radical action or seek professional help. This book can still be your faithful guide and assist but no one should stay in an abusive relationship. These exercises will support you through unraveling the core causes which would lead to your attracting and allowing the situation and are all the more crucial. It's possible that you doing this work for yourself could shift the dynamic and lead to healing for both parties and a further commitment to work together.

The next exercise deals with memory.

Exercise 2: Drawing Out Your Suppressed Pain

This technique releases suppressed emotion through memory.

Time: 15 minutes to however long you need. If you can allow yourself, it can go for some time if the memory is long and convoluted.

Supplies: Paper or a large unlined journal, pencils.

Preparation: Seal the page. Simply say, 'I bless this page.' As energy travels, blessing the page confirms the images and narratives are for release and healing only. Take some long deep breaths.

How to do it

Step 1. Center yourself and scan your body as before. When you feel a sensation in your body breath into it. Allow a memory to emerge. The mind and body are remarkably obedient to suggestion. Trust. If you, like many, have ignored the second brain in your gut, it may take a few sessions to start it firing, but it will. As you connect to it, body intuition will grow and grow. It is possible a metaphoric narrative will emerge. That's fine. It's the language your body is using at the moment. It's also possible a specific memory will emerge, perhaps even one you have completely forgotten. Allow a memory to emerge from the part of the body that is lighting up. Continue breathing into it.

Step 2. Begin to draw this narrative with stick figures as if you are creating a cartoon, drawing out each phase of the story. Stay focused in your body. Keep going. Draw exactly what you feel. Concentrate more on how you feel than accuracy. Children have huge emotions. What did you really feel like? Did you feel hurt, embarrassed, angry, aggrieved, devastated, humiliated? Write the emotions you felt at the time on the paper. You can use thought bubbles if you have something to say or scream.

Once you have drawn out the scenario concentrating on your emotional memory, take a breath. Drink a glass of water. Now return to the drawing.

Step 3. What would you have done if there were no law, or rules of civilization? This is why we bless the paper. This is not about vengeance, it's about release. Go for it. Don't hold back. For example, you may have felt like killing a parent. Or you may have felt like you were being killed. This exercise does not encourage violence; it releases the residue of violence from your system in a safe way. If we were all taught this process, *there would be no actual murders.*

Sometimes there will be a big catharsis, an emotional release of the emotion suppressed at the time, sometimes it will be just literally drawing it out using both meanings of the term – to make an image and to draw out a poison. Write what you would have liked to have said. Don't question it, keep going until you have reached the end. How will you know? Body intuition and how you feel.

Keep checking in on your body. Keep breathing. If you begin to feel disconnected, stop and feel into your body and then keep going. Drawing the memory fires your mirror neurons. It really helps draw up the emotions connected to the memory that are been trapped, causing anxiety, paralysis, triggers etc. Check in to see how you feel after you have expressed your emotions exactly as you would have liked to at the time. Emotions are layered. Rage usually covers grief or sadness.

Step 4. When you have exhausted every detail, draw the ideal scenario. Draw your parent or caretaker (or lover, spouse, employer) treating you with love and respect and you responding with loving authentic communication. This creates new neural pathways. This step is very important. I have found it is human nature to arrive at true forgiveness and compassion after authentic catharsis. All anger and vengeful feelings have been drawn out. Drawing the ideal scenario is a crucial step.

Do this exercise as many times as you need to and then whenever you experience a painful memory. If you are responding to events in your life with huge emotions such as anger, despair, mental stewing, agitation, paranoia, anxiety and so on, the habit or pattern of your response will have been laid down in childhood. You can start with current events but must eventually find your way back to childhood events. It is the first time that the emotion was suppressed that must be healed to free yourself from its ongoing effect on your life.

Whenever you experience a painful reaction to a situation causing you to feel triggered, ask yourself, *When did I last feel this emotion?* And then, *When did I first feel this emotion?* Creative Alchemy is a treasure hunt. It is also very powerful to ask yourself these questions after Exercise 1 and move into the memory work of Exercise 2. You can then go back to painting raw emotion and then again to drawing memory, alternating between the two exercises. In private sessions I guide clients to move back and forth between these exercises for two or three times.

It's advisable to express at least one memory fully per session. This can take a while but please give yourself this gift. When you're finished, you will feel genuinely resolved about your painful memory. Even if you don't reach catharsis you are giving time and attention to your body and emotions and you are creating new neural pathways which will build momentum and shift patterns in your life. You are less and less likely to draw triggering situations to yourself related to this suppressed trauma.

When to use this

Whenever you are feeling overwhelmed by emotion. Whenever you have some spare time to give yourself this very important gift of self-knowledge and clearing.

My Experience with Drawing Out Suppressed Pain

I was outwardly successful: married with a beloved daughter, engaged in an exciting career and living in a wonderful place. We lived above a thriving theater and a beautiful Victorian pub my husband had founded, which we ran together. But inwardly the unreleased trauma of my childhood was summoning the crows home to roost, creating internal confusion and drama that was affecting my marriage and my daughter. I so wanted to give my child everything I had not had myself and was heartbroken that my unhealed self could possibly adversely impact her despite all my efforts, by making me emotionally unavailable at times or melancholic or even secretly despairing, although there were also times when I was buoyant, enthusiastic and positive. I was over-sensitive and easily triggered.

I have always had heightened awareness, and through my travels and studies and training in several healing modalities I had gained many insights, knowledge and tools that I could use to help others, but nothing had quite done the trick for me. I was the classic wounded healer.

Perhaps inspired by the intensity of my inner call for help, I was blessed with a huge life-shifting awakening and a parting of the veil, allowing me to directly perceive the ascended realms, dimensions as near to us as a heartbeat. These dimensions, or octaves as I like to call them, resonate with a frequency so much faster than ours it prevents us from perceiving their existence. Think of a sound so high-pitched you can't hear it, or the blades of a fan spinning so fast you can't see them. Much of the methodology of Creative Alchemy is what I was given to heal myself. This break in the membrane of dimensions (or sudden tuning of my inner 'dial' to the right 'radio station frequency') opened me up to a previously hidden world of palpable love and deep wisdom. One of these techniques, Body Communication, was co-created by my colleague on this journey, Christina Hagman. Our experiences together and this radical awakening are chronicled in my memoir, God's Theory of Creativity: an Odyssey.

Among other techniques received, Exercise 2, Drawing Out Suppressed Pain was one I used continually over a long period of time, approximately two years. This process of setting my emotionally charged memories on paper would sometimes take me hours, but I was diligent. As I drew out the story, I was guided to allow, on the safety of the page, any expression I had suppressed at the time of an event I was describing.

For example, if I was working on releasing a buried memory of my mother's violence, I drew this in stick figures in meticulous detail. I blessed the page so the energy I was releasing was qualified to be only constructive. I then, with all the passion and uncontained fury of a child, drew myself chopping my mother to pieces. I felt a huge burning sensation in my body at the site where the emotional memory had been stored, which in this case was my right torso and my liver. I have dealt with a serious liver illness in my life and I have no doubt that emotional suppression can lead to physical illness. As a facilitator of Creative Alchemy workshops and also one-on-one sessions, I have found that my body and the bodies of others can literally start lighting up like pinball machines when suppressed memories and emotions associated with them are drawn or painted on paper. It's as if the body longs to divulge its secrets and become free of the burden of them. It's as if the memories themselves want to be free.

After I drew myself chopping up my mother I felt terrible remorse. So, I reassembled her – still using simple stick-figure drawings. Then I drew what I had wanted as a child: her holding me and rocking me. I was sobbing by this point. My whole body was burning. I felt deep forgiveness towards her. Not the artificial forgiveness we often force on ourselves when we want to be good people but deeply genuine forgiveness.

By this point in the process, I was able to vision – to clearly see on an intuitive level – what was unhealed in my mother, and understand how that had impacted her ability to mother me. Something very powerful had been released and some of my lost self-worth had been restored. I have witnessed this in others who have given themselves the time to utilize this simple process.

Cassie's Journey

Cassie is a lively, very positive animal therapist. She came to a session to experience something new. Her first painting expressed fear. Her second painting, moving from brushes to hands, expressed anger. Moving back in time to when she first experienced anger, she came across a very charred old part of herself. She arrived at her child-self, who recalled feeling very overlooked and left out. She wanted to run away.

We moved to drawing and she was encouraged to turn around, not run away, and tell her family what she felt. She expressed her anger in several stick-figure drawings with word bubbles containing verbal release. After anger came sadness. Moving back to painting Cassie expressed this sadness. Encouraged to verbally communicate with the sadness, Cassie returned to the charred part of herself and found it had revived. She allowed it to speak. It transpired this was her abandoned genius, an aspect of her multidimensional self that had been left behind long ago. It began to impart new ideas and new of ways of negotiating the world. Her younger self was also present, feeling more relieved and resolved.

Cassie left the session with a sense of wonder about the vastness of our beings, a much higher feeling of self-worth, a commitment not to run away from her feelings and to get to know this previously lost part of herself that had so much potential.

The body thinks in images, so it requires only very simple activities to access, activate, unravel and release buried memories and their accompanying emotions such as painting or by drawing stick figures. When we are immersed in the memory our mirror neurons kick in providing us with the exact emotions we felt at the time of the trauma. Maybe it's a memory of abuse or a parent favoring a sibling over you, or a parent who was continually overly critical, or you being told you were bad if you expressed extreme emotion. The hurt you bottled up is now stuffed up in your heart or the part of your body that got smacked or in your liver where anger hides or in your lungs where grief builds up and so on. What you didn't do at the time was express your true authentic feelings and now you have to, in order to be free. No one is to blame. These patterns have been passed down for generations. You are now the one to break them, heal them and create new ones. Relief will be felt in your entire lineage.

If we shout or stomp away or throw tantrums, that isn't necessarily expressing our feelings. That is a volcanic overflow of suppressed emotion. Genuine emotional expression is accompanied by awareness and articulation. Children need to be taught this or even better have parents who model it. Parents who control their children by smacking or threatening or suppressing are not modelling authentic emotion. To express authentically is to say something hurts and be met with compassion and concern our parents frequently don't have the energy for, or the tools. We have to learn now how it is done.

Often children are given a sweet if they are upset or physically hurting. Expressing how they feel either verbally or with drawings or paint would be much more effective. In fact, sugar can be a factor in tantrums as was the case with a few children. With the support of a naturopath, someone who is an expert on diet, it was revealed their uncontrolled emotion was partly caused by blood sugar crashes that followed excess sugar intake. The crash would catapult them into frightening out of control tantrums. Greatly reducing sugar intake and giving these children Creative Alchemy tools for the expression of their strong but as yet inarticulate feelings turned the situation around. We continue this pattern of avoidance as adults by comfort eating or drinking too much alcohol, taking drugs, impulse shopping, meaningless sexual encounters and so on that leave us feeling spiritually bankrupt and no better the following day, usually worse. The antidote is safe expression of buried emotion.

When you are expressing your emotions to your loved ones, friends or colleagues, start with how you feel and not how they made you feel. Take a breath and say you feel triggered or hurt and ask if you can have a meaningful discussion. This is the hero's journey, when relationships become mirrors for growth. If you are angry you have to own it and not project or blame. Self-responsibility for emotions is the path of mastery. You can say feelings of anger are coming up for you or you are feeling rejected or abandoned, and ask if it can be discussed. Make sure you are also doing the work to help heal yourself and share that if appropriate. Authentic emotional relationship takes enormous courage. We have to let go of false pride and the need to be right no matter what. You may also find that your current partner or friends or work situation will change. Eventually you will have people in your life who can communicate honestly and from the heart most of the time and are able to receive honest communication from others. When you commit to healing your life, you will magnetically draw individuals on the same path. Like attracts like. Friends and loved ones may ask you what you are doing as you will seem brighter, more directed, more confident. You could inspire their healing journey. It takes two to tango and if you've stopped overreacting they may also find new ways of communication.

We are energy beings, *frozen light* as the scientist David Bohm put it, waveforms interfacing with waveforms, connecting with one another, all of nature and the global electromagnetic field. Your new cleared energy field affects *everything* positively.

As we create a new memory of the past when these patterns were first laid down, we create the potentiality of a new future. The connections between neurons literally change in our brains. When you are healed of suppressed emotion you will still feel deeply. Life happens. Relationships break up. Jobs dissolve. When you are truly in the flow, it is easier to see these sorts of events as avenues to new opportunities while still allowing and honoring your natural emotional expression.

When to use this

When there are signs of suppressed emotion. When you become aware of unhealthy patterns repeating themselves. When you just feel out of alignment or if you are physically unwell. Let the illness speak to you through images and words. Trust anything that comes.

The session below may seem a bit 'out there' but it was very real to experience and caused lasting change.

Malcolm's Journey

Malcolm is a clinical hypnotherapist who is on an accelerated journey to personal mastery to benefit his clients and his own true purpose. His first drawing expressed a range of emotions from fear to grief, caused by a memory that immediately surfaced of some of his family members witnessing him struggle against strong currents in a river. His younger self felt sure they would have been happy to see him die. A second painting revealed a dark form hovering near his crib as a child. Moving into the image he felt the figure was an entity which had followed him from another life.

Moving Malcolm back in time, he recalled a life as a young, not very advanced practitioner of the occult in medieval France. He had called in a helper from the lower astral realm who had stayed with him and who, in accordance with its nature, had fed on him, even veiling the perception of situations in his life to cause fear, its source of sustenance. Malcolm drew out this memory and also wrote what he felt about being fed on and 'haunted.' Together we called on the Great Ones to take this being to the higher realms for training so that it could evolve and free Malcolm.

Malcolm, who is partly of Māori descent, then felt something troubling in his stomach. It was revealed as a very unwell Māori brain. He could see and feel the amazing way Māori culture felt and perceived the land, the oneness with it and heightened perception of its beauty and what a huge gift that was. But he also felt the sickness caused by colonization that had coalesced into this brain in his gut. I energetically removed the brain and asked Malcolm what he would like to replace it with, suggesting a 12th-dimensional Māori brain. The indigenous culture of New Zealand is very connected to Earth and also galactic wisdom. He chose this. Carefully, I lowered the 12th-dimensional Māori brain into Malcolm.

The Archangel Uriel appeared and stated he would overlight Malcolm. A celebration then occurred and Malcolm was invited to stand with the Guides, those who assist us on our path to freedom, in honor of his deep commitment to heal others and raise Earth. He was given various rays as gifts, which suggests the development of the Rainbow Body.

After the session he looked very different. He felt radiant with life. The amazing thing is that little miracles keep happening with regards to the 12th-dimensional Māori brain. People near him can suddenly speak fluent te reo, the indigenous language, or can converse with the Māori spirit guardians who sometimes come to his garden. The important thing is something has been freed and Malcolm has come into greater balance, sovereignty and self-worth.

What to Expect from This Part of the Programme

As you use these techniques to cleanse your emotions and release the stories in your body you will start to rehabilitate your nervous system, laying down new neural pathways, releasing beneficial neuropeptides that make it easier and easier. You will have an exhilarating free-flow of emotions, feeling deeply as they pass through you without getting stuck.

Free-flowing feelings enrich our lives. Sorrow. Joy. Anger. Grief. Excitement. Happiness. Joy. Love. All emotions are part of the rich fabric of our lives. Later we will learn how important strong feelings are in creating our radiant lives and how to apply them as fuel.

When we have a way to express and our emotions don't get stuck, we can feel deeply and we feel very alive. Gradually anything we do to suppress our emotions or thoughts becomes distasteful. The more we clear ourselves out, the less we will experience depression, melancholy, hopelessness, lethargy or anxiety, and the less we will be triggered by people and situations. If you feel antsy, stressed or angry, or you are going through life as though everything is okay but your heart is screaming, this is where you want to start. This is caring for *you* on a deep level. Love in action! We *all* have suppressed emotion.

The more of the old stuff you clear out, the more life force you will have. This released life force activates your inner superpowers. It aligns you with your I AM Presence simply by removing the barriers. The depletion of life force caused by lifelong suppression causes aging, lethargy, depression, anxiety and decline in health. Increased life force is a magnet for more abundance, creates more youthful vitality, inspiration and 'aha' moments of radiant clarity and joyfulness.

The simple techniques of Creative Alchemy allow a magical alchemy to occur, the creative process by which one substance is transmuted into another. The stored emotion with its attendant feelings flows outwards and changes from lead to gold. There is an enormous sense of relief. Many people start with a strong emotion and find halfway through it is already transforming into another emotion. Practice changes the brain's structure and function. It all gets easier. It becomes fun!

More Painting Emotion and Stick Drawing Stories

Brandon expresses the grief and humiliation he felt from his father's response to his enormous sensitivity, a continual litany of remarks like 'Buck up' and 'Be a man.' He finds relief and the ability to express vulnerability and emotion drawing out these memories, alternating with powerful emotional releases using paint. The release of emotion increases his alignment and insight. He can now see how frightened his father was. As the armor built up over years of suppression falls away, his ability to experience emotional intimacy with his life partner increases. He is now able to see his sensitivity as an empathic gift and this has influenced his career choices.

Jason, who was continually beaten as a child, draws several scenes between himself and his mother, who was alcoholic. He first draws throwing his mother off a mountain, expressing rage that looks like fire pouring from his mouth. He then draws himself as a child having feelings of huge remorse, a small figure hunched over with tears coming from his eyes. These two pages help him uncork a lot of bottled up emotion. He misses his mother and begins to experience the beginnings of compassion for her. He then draws a third image of his mother holding him. In the following weeks, he experiences a significant decrease in anxiety and chronic back pain and initiates contact with his mother after several years. He finds he is able to have more compassion for her and is also better able to perceive her love for him and how her own wounds had caused her then uncontrollable behavior.

June experienced sexual abuse as a child. Over several sessions, using the steps of drawing the scenes of the incidents that wounded her and then expressing verbally by writing how she really felt at the time of the violation, she releases shame, horror and grief. In successive paintings, she begins to release rage that has been stored in her womb and a feeling of paralysis in her throat. Very painful menstrual cycles begin to alleviate. Her normally hesitant way of speaking becomes more confident. June finds herself more inspired in her daily life and has increased innovation and patience with problem solving. The shame and grief that made her feel frozen has gone, but she still feels angry and unforgiving toward her abuser. She is committed to continuing until she can release those feelings and free herself.

Feelings of anger and vengeance may persist, and although they are commonly seen as natural in our society, we are only freed when we no longer drink the poison of our own wrath. Perhaps a better word for *forgive* is *release*. It's also important to remember we have an emotional, physical, mental and energy body. Some people can claim forgiveness mentally and emotionally, but still carry rage physically and in the subtle energy body. Remembering to breathe deeply all the way to your core and without ceasing, in long slow circular breaths (not hyperventilating!) ensures you are reaching all four aspects of your earthly vehicle.

This process should never be forced. Everything must be handled in its own time. Last time we spoke, June told me she had joined a group of women who have also experienced sexual abuse and help one another in many ways. This is something she never considered doing when her wound was hidden and secret. She is also experiencing light-heartedness and more trust in the world.

As we are relieved of these buried memories, we are less wed to the *stories* of our emotions. We stop the inner, 'this happened and that happened and he or she did this or that.' We find we are able to experience our emotions purely. We can drop into anger and experience the pure power and fire of it. We can sense how this emotion can become passion and compassion. We can feel sadness and not cover it up or drown it in distraction. At this point we may want to find different forms of creative expression, for all authentic creativity helps keep emotions flowing. Time to take guitar lessons? Start painting for the pure pleasure of it? Or writing? Walk on the earth with bare feet. Dance!

Exercise 3: Quantum Communication

This exercise clears trauma through writing.

Time: As long as it takes. Ideally begin when you have at least a few hours. That will also ensure you don't feel time pressure.

Supplies: A pen and a journal or paper.

Preparation: Center yourself and take some deep breaths. Bless the page.

How to do it

Step 1. Choose a significant authority figure from your childhood. Mother, father or significant caretaker. Some people write a letter to God.

Step 2. Write a 'no holds barred' letter. Take your time. Everything you ever wanted to say and never said. Everything. All hurts, slights and misunderstandings, blames and projections. Let it rip. If positive memories flow, write them as well but don't cover up emotion or write positively to assuage guilt. This is an emotional clearing exercise. Keep going until you are absolutely finished. Avoid reading the letter after you are finished and until the second half is completed. If you do, you'll engage the analytical mind which will stop the flow of emotion.

Step 3. Now, ideally directly afterwards if possible, write a letter back from that person. You just take a deep breath and dive in and let the quantum field do the work. If you weren't raised by your parents, then significant caretakers. It could also be a sibling, a teacher, ex-partner, ex-boss, etc. Whoever looms from your childhood and who you have blamed.

I used this technique with suicidal patients as part of the 12-week trial at the Stress Project in London in the mid-1990s. The Stress Project was set up to offer several alternative modalities simultaneously to patients under the care of doctors and psychologists. I was asked to trial Creative Alchemy, then called Source Alignment and I created a 12-week programme to do so. The first six weeks were used for clearing and the second six weeks were building self-esteem, emotional resilience and the desire to live. I presented this exercise. The first part, writing a letter to the person that they perceived had persecuted them in some way in their childhoods, went well. Then I presented the second part, which was to write a letter back from the person who they felt had harmed them, abandoned, neglected or rejected them or, in many cases, severely abused them. At first there was huge resistance to the return letter. Everyone expressed disbelief in its efficacy or even the possibility of writing from another person's consciousness. Horror at the thought of connecting this way with parents or caregivers who had in many cases brutally abused them was loudly exclaimed.

Everyone in the group of 15 people managed to write both the letters to and the letters back. Most of them accessed profound messages from their parents which astonished us all. Our Higher Minds are connected by the quantum field or the Collective Unconscious as psychologist Carl Jung called it. The return letters expressed insights and awareness that seemed inexplicable. For example, explaining they did not know how to love, they had done their best, they felt remorse, sometimes with specific details from their own childhood, all to the astonishment of the patients. Some parents 'wrote' back that they had been abused themselves. All 15 patients experienced greater understanding of the river of the unhealed repeating itself and also experienced greater compassion for themselves and their parents.

You owe it to yourself to give yourself the time and the gift of this exercise. Quite astonishing shifts can occur with this technique. Without ever communicating directly, the inner relationships have been healed even if it may not be possible in the outer. Terrible actions have been forgiven and released and the poisons transmuted.

That parent or once-towering figure from our young life could play away at an old routine that used to send us into paroxysms of rage or grief or frustration and now we just give them a hug, ask if they want a cup of tea, change the subject. It always takes two to tango and when there are no longer two engaging, old patterns die off. Often, as we transform ourselves, altering our frequency and the energy we are radiating and aligning more with our true nature, we find that the people in our lives change on a deeper level as well. This is partly because we are much less triggered by our trauma or resentment or fear from the past but this often occurs even if there is geographical distance. The radiation of our energy fields defies time and space.

When to use this: This exercise is important to do at least once for each parent. It is also useful whenever you are in a significant relationship and communication lines are down.

If there is a lot of hurt and rage, a very powerful act is to create a ritual and burn the letters. Make sure it is a safe fire! Afterwards gather the ashes and bury them, or place them in running water, the sea, a river or even the kitchen sink. This powerful activity helps transmute the pain on a deep energetic level.

Many people report significant changes in these relationships and most importantly, within themselves.

Exercise 4: Befriending Emotion

Time: As long as it takes.

Supplies: None.

Preparation: Willingness.

This is a technique for handling full-blown emotion when it hits like a tsunami. The heart of the technique is training yourself to feel your emotion. That probably sounds like a contradiction in terms. The fact is that we are usually too busy mulling over the story of how we feel the emotion came to be to actually deeply and sincerely experience it. This is actually a coping ploy to keep us in our heads and out of our bodies, so we can avoid the uncomfortable feelings.

How to do it

Step 1. Next time you have an overwhelming emotion, check in with your body. Stop, listen. Take a seat or lie down if you can. Feel your feet on the ground and your bottom on the chair or your body on the bed or the grass. Get back into your body and out of your head. Rather than blame, project or drop into the repetitive story of why you feel as you do and what has been 'done' to you, *just feel.* Stay there. If you wander back to your mind, come back to the body and the experience of pure feeling. Let the pain or anger or grief or frustration or anxiety just be. Immediately take off the label of good or bad.

Step 2. Remind yourself that you are a creator. Stop sending darts to whatever or whoever seemingly caused you to feel as you do and take full responsibility for creating the feeling inside yourself and *feel* it. It's amazing.

Transforming an Emotion by Experiencing it Fully

I was feeling enormous anger. Someone whose opinion I cared about deeply had mocked me and had been very cruel. I felt that I was being misunderstood and betrayed and I was very aware that this experience was a pattern for me, one I created through my perceptions and beliefs, but I was unable to stop blaming my friend for how I felt. I felt shocked and unloved in the extreme. I knew that to project responsibility outside me – to perceive that something was being 'done to me' – was to take the position of a victim. In effect, feeling victimized was a sign that I was no longer master of my life. This time I decided to take responsibility for drawing this situation to me, like a magnet, based on what remained unhealed within my psyche, and more importantly, to activate my ability to choose my reaction.

With a huge effort, I pulled myself away from the story that my mind was repeating over and over and finally succeeded in disengaging from it entirely. The entire scenario burned away, leaving just the emotion, which was grief. I breathed deeply into the pure grief and found anger underneath it. In allowing myself to be present in my body and fully feel the sensations of the emotion, something shifted and I began to feel the burning fire of my rage as pure, vital, astounding energy. I allowed this force to move in me and through me, overwhelmed by its power. My whole body felt on fire, but the feeling was overwhelmingly positive. I felt electric and very alive. I wondered, even as I allowed myself to be consumed by it – this pure fiery energy, is this really me? This much pure power?

At last, I realized that this fiery rage was also the energy that transmuted into the strength of my passion and that it also held within it the softer embers of my compassion. Righteous anger can also be a catalyst for change and justice, or a healthy physiological signal our boundaries are being transgressed. How could I give this away to another by blaming and projecting, spilling this precious fire, harming them and losing its energy and value for myself? How would my younger life have been different if I had learned to own this power sooner?

If, while trying this, you feel too overwhelmed, then move with it. Dance it out. Sometimes when we are authentically experiencing our emotions for the first time it is almost too much to bear. So many of us are in our heads and not even near fully embodied. We are now dealing with two new things at once, the experience of an emotion and the richness of the experience of being in our bodies. For me, this exercise makes real the poet Rumi's insight, *the wound is where the light enters.*[164]

When to use this

Do this exercise any time you feel caught in a strong emotion.

Exercise 5: Quantum Body Communication

Directly communicating with the body.

Time: 15–30 minutes.

Position: Comfortably seated.

Supplies: Pen and paper.

Preparation: Center yourself and take some deep breaths.

How to do it

Step 1.   Take some deep breaths and allow your body to point out an area to focus on. Scan your body feeling into the sensations.

Step 2.   Now start to write. *What does that body part have to say?* If you are having physical issues communicate directly with that body part.

With this technique, you will directly experience that consciousness is held in all parts of our being. The body is our book of life. Communicating directly with your physical body allows it to express profound and inspiring messages of what you are neglecting, what you need to change. The body has its own inherent wisdom. Ask what it advises. Life style changes? Sometimes the advice is profound and beyond personal knowledge.

When to use this

Do this exercise any time you have a physical issue or illness. Do it at least a few times just to start a dialogue between your physical body and your mind.

You can also have one-to-one communication with your emotions. Your sadness can speak to you, or your fear or your grief. These can often be quite beautiful and have special messages for us. Everything on Earth has consciousness and can be communicated with and desires to communicate – animals, plants, rocks and people. If we were taught this from childhood, we would treat one another, our selves and our planet very differently.

Examples of Quantum Body Communication

Raina, a 65-year-old engineer, communicates with her left ear, which has been deaf for decades. She is profoundly moved and receives messages for its healing and also its grief from being ignored and overlooked.

Robert, a successful entrepreneur, receives messages from his diseased liver about what he needs to do to heal it. He learns about changes he needs to make in his life.

A six-year-old boy is told by his stomach to stop eating so much sugar.

Several teachers all get similar messages from their backs, which have been troubling them – this issue seems to be common among teachers. Some received detailed instructions for what they could do to relieve the pain, referring to treatments for adrenal fatigue and other medical conditions that can cause lower backache, and also including dietary recommendations, movement disciplines and sometimes a change of career.

I've had profound messages from many of my organs, including a strange one where I received the image of stags gathering around small yellow flowers. I looked up the flowers and learned they were St John's Wort. I knew a tincture of it could be used for depression but I found they had another use that was pertinent to my issue. St John's Wort is also antiviral.

Exercise 6: Accelerated Body Communication

Special thanks to Christina Hagman for co-creating this exercise with me. We facilitated each other through powerful healing catharses the summer of 1995 with this exercise and it has proved very powerful for others as well. It's wonderful if a colleague on the path can be with you and you can take turns alternating, but it can also be done on your own.

Time: 20 minutes to an hour.

Position: Lying down.

Supplies: Journal and pen.

Preparation: Willingness.

How to do it

Step 1. Lie down and begin to breathe deeply into your body.

Step 2. Become very aware of any part of your body that is drawing your attention. It can be very subtle. Breathe into your body while scanning it with your mind. Either by sensing, feeling or imaging, stop where you sense, feel or see energy. Energy is always an information carrier. It could be the sensation of a small irritation or any sensation at all. It could feel like a little signal. It could look like a light on a pinball machine. When the part of the body that presents itself begins to feel 'activated,' start to breathe slow deep breaths, which will begin to push up and out any buried emotion. As long as you feel the charge, don't stop breathing. Make long deep circular breaths so there is fluidity and connection between the in-breath and the out-breath. Keep breathing until you experience catharsis.

Step 3. Let it out using movement. Have some pillows nearby to throw or bash. Roar. Scream. Shout. Make discordant sounds loudly. Jump up and down. Lie on your back and kick your legs.

Step 4. Write about the memory you've seen, felt or sensed in your journal.

When to use this

When you feel so much build-up of suppressed emotion you want to blow the lid off. This can happen after using the first two exercises and all the buried emotions start to arise. This is a great exercise for removing impacted emotion from the physical body.

An Example of Accelerated Body Communication

Isabel, a young woman with kidney cancer, retrieves a memory of her brother's suicide hidden in her kidney. Her family had completely shut down at the time of her brother's death. All photographs had been removed and no one spoke of him. There had been very little expression of grief or honoring his memory. Upon accessing this memory, Isabel begins to cry. She also says she feels extreme heat in her body in the area of her kidneys. She starts to breathe into the heat, causing it to expand. She begins to sob huge gasping sobs. Finally, she stops and goes into stillness. She receives insights into the extreme grief that created the behavior and she feels the presence of her brother and senses he is all right. Months later she reports the tumour has broken up and is dissolving against all medical expectation. (PLEASE DON'T STOP MEDICAL CARE. CREATIVE ALCHEMY IS AN ALLY, NOT A SUBSTITUTE.)

Marta, a 40-year-old woman tortured by paranoia and anxiety, accesses memories of childhood sexual abuse that she has hidden in her lungs. She imagines herself killing her perpetrators and starts flinging and bashing pillows. She screams the scream that has been buried in her lungs for 37 years. After expressing rage, she begins to cry. She remembers the last time she felt she was beautiful was as a little girl before this rape. She uses emotional painting to express her grief. She begins to experience relief from her symptoms and renewed self-esteem.

Tomoaki, a 60-year-old scientist, accesses memories of being taken from his mother as a small child. These have been buried in his left shoulder and his heart. He has an emotional catharsis. He has not trusted women or relationships or himself, his entire life. Chronic anxiety begins to lift.

Geena, a young woman who had lost her father as a toddler, finds her grief stored in her chest and allows it to pour out with a massive roar, in the process relieving her chronic asthma. Over the next few months she continues to have Creative Alchemy sessions. More of her trauma is released and she gains insights into her fear and distrust of men that begins a shift in her perspective.

With this technique we use breath to access where in the book of life, aka the body, a suppressed memory is stored with the intent of cathartic release through physical movement and vocalization. Sound is powerful. It shatters what is frozen. As children, we vocalize continually and then shut down this function as adults. Kicking our legs and other childlike actions can help release societal conditioning and allow our sounding and actions to be expressed in an extreme cathartic way in a safe setting.

Your conscious awareness is becoming heightened by the earlier exercises and will be able to guide you to where the suppressed emotion or trauma is stored in your body. Your body wants to cooperate. Using guided meditation and breath, our bodies will literally light up calling attention in our minds. These memories may first come as stories or metaphors and will gradually reveal the emotions and memory of events where we sealed our lips and did not have the tools to authentically express our feelings without blame or projection. Rather than draw or paint them we are going to physically move them out and allow extreme sounds and actions to accompany their expression.

Studies done by German surgeon Ryke Geerd Hamer MD, with hundreds of cancer patients, revealed that they had all experienced a big shock a few years before the cancer appeared. He later created a more comprehensive study surveying 31,000 patients, which logically and empirically revealed that all disease is initiated by biological conflict, shock or trauma.[165] He discovered that this trauma leaves a visible mark on the brain that can be confirmed by CT scans. The scans furthermore confirm that trauma marks the brain in the area that relates to the functioning of the organ or body part affected.[166] His theory of why this happens is that during a time of severe emotional distress nature intervenes and shifts distress from the psyche to a corresponding body part, allowing the individual to remain functional and not collapse under the stress.[167] Given these dramatic findings, it stands to reason that the trauma must be relieved or the organ or body part involved will be aggravated into illness.

Hamer went on to create a complex chart that aligned various traumas to certain organs or body parts. By identifying the trauma and releasing its emotion we can relieve the organ or body part of stress and create new synapses in the brain, allowing a new present and a new future. In some cases, illness is too advanced to shift or is part of the life journey of the individual. It is still important to release the buried emotion. This will affect their transition positively.

The Practical Application of Creative Alchemy now shifts into building your muscles of vision and transcendence to prepare for Phase Two.

Exercise 7: Rewriting the Script of Your Life

Supplies: Pen and paper or journal.

Time: As long as it takes. It could take weeks off and on or could be done on an away weekend. It is well worth giving yourself this gift.

Position: Comfortably seated.

Preparation: Center yourself and breathe deeply. You are going to breathe in a different way now as you are not accessing or releasing buried emotion. You want to connect to the deepest part of yourself. Breathe long deep breaths all the way into your root chakra. Feel the muscles relax and drop. Breathe it up again to your heart. Now breathe up to your crown. Breathe back and forth this way, very long slow breaths up the back of your spine and down the front of your body. Put the tip of your tongue on the palate of your mouth just behind your front teeth helping to create a circuit. Slowly and gently repeat this breath for a few minutes, finally resting in your heart.

How to do it

In this technique you will rewrite your entire life script from before birth, all the way through to when you step out of your mortal coil for another grand adventure. It is your chance to envision all the circumstances that would give you a sublime life. And it is your chance to correct every perceived wrong and remember every event at its optimum, establishing new neural pathways and transforming the responses of your mirror neurons. You can also choose to write up to the present moment and save the second half for another time or after you have completed all the exercises in this book, as your dream may well change. It can be worth doing the second half now as a record and doing it again later. Giving yourself this time and loving attention is invaluable.

Step 1. See this exercise as writing an exquisite love letter to yourself. You can make a book of it and fill it with images. Images you love from your life, new images where they are needed and future images for what you want to create. Start your story before birth. See your free, radiant, extraordinary spirit deciding on what courses you want to take during your precious time at Earth university. Now begin to write the most wonderful story. Begin with your Spirit scanning Earth for the perfect parents to learn what you chose to learn in this life. Your beautiful gestation and how soft and safe it was. The music your mother plays, the cooing sounds of your mother and father, your perfect birth, then on through your childhood in enormous detail. Your pony, the fields and streams you played in. The great school, big family, many friends. Whatever floats *your* boat. Right up until now and then keep going if you choose. Chart out your perfect life from now on until you decide exactly how and when you will exit your quantum navigator for your next big adventure. Give yourself a long life! See your complete transition from Homo sapiens into Homo illuminatus. You are healthy and fit and abundant and loved, with a wonderful meaningful life purpose.

You are not being false. You are using your power as a conscious creator to create a new you. Allow yourself to feel all the accompanying emotions. Liberally apply gratitude. Your mirror neurons will be recording this as real and your brain chemistry will be responding.

Step 2. Finish your story with a list of wonderful things about yourself. People often resist this. We've all been told, 'don't boast,' but so many suffer from low self-esteem. How about we each become the best we can with no completion? Create an exquisite soul poetry, a litany of beautiful qualities and potentials. Give yourself this gift and be honest. You will have at least one full page of wonderful qualities and potentials at the end of writing.

Step 3. When you are finished get a beautiful folder to keep your story and images in (if you choose to have images). Or you can get it bound. Look through it every now and then and keep it in a place where you see it often. If you want to make changes, do so.

Do not rush this exercise. Fold it into your life dedicating what spare time you can until it is done. You are worth it! To shift your inner imagery and lay down new neural pathways, you need to take the time to do this exercise slowly and with consideration. It is best if it is very detailed. If it takes weeks or even months, that's fine. You are creating a brand-new way of being. Invest in yourself! As you do this you will be elevating your frequency – your soul song, your tone, your blueprint – which will change the quality of what you attract in your life and improve your health on every level.

You will notice your feelings about your past, present and future change. The pre-existing script of your life has been hidden in your unconscious mind where it plays like a broken record, creating your current results. Repetition of your new script will help you to solidify the new results you long to see. Albert Einstein stated all time is happening all at once and that it is circular and not linear, something the mystics have always known. If we send waves of qualified energy down the various timelines and dimensions, we can appear as angels to our younger selves, members of our ancestral lineage, our progeny, our community, country and world simply through focus and intention and powerfully radiating compassion, healing and love.

Consider recording yourself reading your new life script. You can then play it back to yourself from time to time. The Theta brainwave state we enter just before sleep is an excellent time to lay down new neural pathways. Gradually, from tumult, drama and suppression we arrive at the shores of grace, surrounded by loving friends and family, living in beauty and abundance, doing inspired and meaningful work that contributes to our highest good and the highest good of all. You are becoming your ideal parent, learning to love, care and provide for yourself, healing all wounds to allow your highest expression of life to flow through.

Thanks to neuroscience, we now know that our brains are malleable, meaning they can be recreated to experience a new reality from new perspectives. We know from quantum physics that what we call reality is responsive and malleable as well. As the mystics always said, this life is a dream, a very real one and a fluid one.

Your body thinks in images, so use your imagination to flesh out the events you're scripting and *use your emotions* to give the new images texture and depth. At every juncture, pause to feel the wonderful emotions always sealed by gratitude, the signature of receiving.

Examples of Rewriting the Script of Your Life

Andrew had been given away at birth with no explanation. He was raised by relatives and was only vaguely in touch with his parents on an infrequent and formal basis. As a result, he suffered deeply from a sense of rejection and abandonment. In this session, he was first given the opportunity to verbally say to his parents all the unsaid words and feelings he had, which flooded out with deep gasps and sobs, with me and a colleague standing in for his parents. He was then helped to verbally rewrite the script of his life with me and my colleague responding as ideal parents would. After just one session, he was able to drop the story of his victimhood, forgive his parents and access his voice. He spoke to a large group, telling his old story and his new story, something he could never have done a few hours earlier. Andrew has gone from strength to strength.

Josie was afraid to follow her dream and was stuck in a well-paying job, feeling guilty that she didn't appreciate it. She committed to the first steps of Creative Alchemy and after several sessions had released buried grief over growing up with parents in a dysfunctional marriage and the repetition of the same scenario in her own marriage. She spent evenings writing, drawing, painting and collaging the new story of her life. She became increasingly elated by the possibilities in life and found the courage to quit her job and retrain as a landscape gardener. She also encouraged her husband to seek more ways to express his dreams. They are doing well.

Learning How to Breathe

Breathing in certain ways has a grounding and calming effect on the body. When we breathe slowly and deeply our heart rate changes. Calm breathing slightly emphasizes the outbreath. Panicked breathing is more gasping and quicker, higher in the body, more shallow and favors the inbreath. If you can practice moving your breath deeply into your abdomen, feeling it swell way below our usual short upper lung breath, you will find it is easier to stay in the parasympathetic nervous system, the calm one, instead of the sympathetic jumpy fight or flight one. Deep breathing also loosens suppressed emotion.

Soften your lower body and let go of all muscular tension. Feel your perineum drop and all your muscles unclench. Take a breath in and let it pour out of you in a long audible sigh. Take another breath and really feel your lower belly expand. Eventually even your sides should bellow out. See if you can take a long slow breath right down to your root. Use your mind to direct the breath. This sort of breathing oxygenates your whole body, invigorates your brain and causes your lower muscles to move a bit like a trampoline. It should feel very pleasurable. In time, you will breathe this way all the time with increasing benefits.

Summing Up – The Technology of Emotion

Our feelings are the poetry of our lives and the fuel of our manifestation, our superpower! They give us our depth and richness and appreciation for life and allow us to feel *into* music and art, the stories of others and the heart-wrenching beauty of life itself. The deeper our feelings are felt, the richer our experience. We are learning how to feel them and not get stuck in them. We want to honor and express them. We also want to express them safely, so they don't harm our relationships or ourselves.

- It is important to experience the full spectrum of our emotions because they add color to our lives and make us who we are. They have a lot to tell us and to teach us. As we slowly release the volcano of unexpressed emotion, they will stop popping out and scalding us or our loved ones! But they will still come and go, flowing like a river through all we do. When we can really experience them, we become deeper people. We are not darting about on the surface, disconnected from our true selves or floating above disassociated.

- At times the processes in Phase One may feel as though you are delving into the depths of aspects of yourself which you would rather not see. It may be uncomfortable, but stick with it. You will no longer continue to magnetize triggering situations to yourself that require these forgotten and long-ignored parts of you to be seen and healed.

- As we clear, we will gradually become integrated with parts inside us that got lost among the suppressed emotions. Look inside yourself for the child who leaps for joy, has no inhibition and jumps into new things with both feet without fear of failure; or the teenager who has to dance whenever music plays and so on.

- Whenever we heal ourselves, we release another layer of sticky, dark suppressed emotion and the light in the world goes up a notch. As energy beings, when our vibrational frequency is higher and lighter, we automatically give permission to those around us to shine more brightly. We are laying down new patterns for humanity.

- You will know you are ready to begin the practices of conscious manifestation when you no longer feel a victim, when you know you are responsible for your life, your choices, your actions and your thoughts and emotions. You are ready to move to the next phase when you generally feel positive and grateful and your negative mind chatter has ceased or has quieted considerably and you are free of doubt, regret and anxiety most of the time. You won't feel so inclined to comfort eat, drink, self-medicate and generally engage in distracting suppressive activity.

- You may find that progress can accelerate and then suddenly plateau or dive. This can be demoralizing but it is all part of the journey. Something unhealed and buried is raising its hand to be recognized.

- As we utilize all the techniques in the Practical Application of Creative Alchemy, our four earthly bodies gradually become cleansed. If we can approach the process with compassion, we can begin to see the setbacks as offerings. They are clues to what needs attention.

In Phase Two you will learn how to develop your imagination and the techniques for elevating it to Living Vision, an important step in conscious creativity.

PHASE TWO:

LIVING VISION

Where there is no vision the people perish.
– Proverbs 29:18

In Phase Two we will delve deeper into your quantum manifestor, your visionary self. The imagination is the province of the mental body, the architect. When you use alchemy to transmute imagination into Living Vision, you enter the province of the Higher Mind, the bridge to the power of the Presence. This is the key to unlocking the imaginal realms, the kingdom of subtle bodies and archetypal intelligences where human and divine imagination meet, mediating between the world of the Spirit and the body.

When we contemplate purpose from this vantage place we have access to your hidden treasure, your 'pearl of great price' – what Hindus call the Ātman, Quakers the Inner Light and Christians the Kingdom. Your visionary third eye opens and you can see beyond the limits you have imposed upon yourself. When vision is activated by the alchemical science of the heart, the Threefold Flame of love, wisdom and power where our Presence is anchored in our earthly bodies, it becomes Living Vision, animated by the same great life force and power that animates your own good self.

Living Vision is differentiated from idle musing or fantasizing, which has no power behind it and isn't intended to manifest (although you should know that recurrent fantasies and musing can gain power through repetition especially if emotional).

You might get to this point and wonder what it is you truly want to manifest. You may have started to question your perceptions and beliefs, which begins to happen as we peel away our culture-scape – the familial and social influences. A lot of buried emotion has been released that formerly colored how you see the world. Suddenly you no longer want what you thought you wanted! Our deepest calling will emerge when we dare to become our authentic selves and express our gifts to the world while fulfilling our own lives. A good question to ask is what did you love before grown-up responsibilities set you on a path prescribed by forces external to you?

When we are young the superpower known as the *imagination* is at our fingertips. We can play all day with just our imagination and some friends. Or some imaginary friends! But then comes that time we're supposed to grow up.

We're meant to buckle down, get through school with good grades and decide what we want to do for the rest of our lives. Go to university and/or get a job. That is if we are living in a culture that provides education. How much say do we have in what we end up doing?

Bringing back a sense of play is hugely beneficial. There is no failure here, just learning. The more you can allow yourself to feel playful with all of this, the easier and more joyful the process will be.

By activating our visionary imagination, we are preparing our vehicle for conscious manifestation. The imagination is not only the architect, it is the problem-solver and the portal to infinity. Elevating the imagination into Living Vision is humanity's great leap and the single most important step in our transition from Homo sapiens to Homo illuminatus.

> *Thou canst not behold me with thy two outer eyes;*
> *I have given thee an eye divine.*
> – Upanishad

Your first assignment will be to build a sanctuary. But before we begin we are going to engage in a meditation that will help you activate your imagination while grounding you and help bring you into a Theta brainwave state. Like a tree, your roots should reach into Earth while your branches shimmer with light from the Sun and the sentient radiation of the cosmos.

You'll begin every exercise in Phase Two with a meditation that connects you to the core of our mother Earth and the core of the Sun. Think of it as a morning ritual or a precursor to doing anything creative that clears the field around you and connects you to your Presence while qualifying the energy you draw from it with love, wisdom and power.

The practice can be immensely powerful for orienting you, grounding you, uplifting you, connecting you. It is a place to avow the power you hold, the love that is necessary to balance it and the illumination that results from this balance. You can record the meditation that is ideal in the beginning until it becomes part of you.

It's ideal to create a dedicated private space you can return to where some of your favorite things can be laid out. You are building a force field just like the high-frequency energy build-up at sacred sites, which can now be measured by electromagnetic sensitive devices. The key to the door is frequency, in particular the Theta brainwave state. Theta is a state of relaxed but alert dreaming.

Achieving this can be assisted by meditation, breathing, authentic creativity and intention. When you have achieved the state a number of times, relaxation, intention and attention are generally all that will be necessary to re-enter.

Exercise 8: Creative Alchemy Meditation

Time: Around ten minutes.

Position: Standing then sitting (or you can continue to stand).

Supplies: A comfortable chair, a compass. Smartphones have
downloadable apps.

How to do it

Step 1.   First, loosen up. Stand with your feet planted wide. Place your hands
on your hips, and move your hips in a gentle circle, going first to
the right and then changing direction and circling to the left. Move
your neck in a gentle circle, going first to the right and then the left.
Enjoy the luxury of this movement.

   If you need support to balance, hold onto the back of a chair or
put a hand on the wall. Next, keeping your toes touching the floor,
lift up the heel of your right foot and make circles with your ankle,
going first in one direction and then the other. Repeat the process
with your left foot, circling the ankle in both directions. Find ways
to move your free leg, which creates movement in your hip joint.
This could be making circles with your pointed foot on the floor.
Or it could be by lifting your knee and making circles in the air.
Never force anything. It should feel delicious to get the synovial fluid
moving in your joints and the flow of energy opening up.

   Circle your shoulders forward, up, and back, a few times. Reverse
the circle back, up, and forward. Then, make wide arm circles – one
arm at a time – circling first clockwise and then counter-clockwise.

   Shake out your hands gently, letting them be floppy at the wrists.
Shake from side to side and up and down. Then shake the arms from
the shoulder. Let the shaking travel. Shake out your torso and hips.
Shake out each leg and foot (hold on for balance if you need to). And
shake everything all at once. This should feel good. You should feel
awake and embodied. Give thanks to your amazing body.

Step 2.   Start to take slow even breaths. Four counts in, hold for a count,
four count out, hold for a count. Imagine these breaths moving in
and out of your heart. Place your hands over your heart to assist this.
Bring to mind someone or something you truly appreciate or love
that gives you a warm feeling. Let this expand through your heart.

Step 3.   Give thanks for all who guide you, and know there are many unseen
beings who are devoted to your fulfilment and are on your team.

Call for guides of the highest, finest consciousness to assist you. You can also ask an ascended master to enfold your four earthly vehicles and imbue you with their qualities.

Step 4. Take a deep breath and feel the breath circulate throughout your body. Send the breath up and through your Crown up to the Sun, and the spiritual Sun behind it, the Presence of our Sun, called the Great Central Sun. Take another deep breath and send it back up through the Crown and connect with your own radiant Sun, your I AM Presence, the source of your life, consciousness and supply. This is anchored in the chamber of your heart where the Threefold Flame blazes. Commit to use your energy with purity of intent and to qualify all of the life force you draw forth into manifestation for the highest good. Fill yourself with cosmic light of the Presence. Imagine your Presence as a sun radiating from your heart, flooding you with warm light.

Step 5. Flood this light through the blue plume of the Threefold Flame, the plume of true power and negative polarity (negative as in magnetic poles). See that blue plume come alive and magnify, bathing your four earthly bodies and your entire force field.

Step 6. Now, flood the light through the plume of love and positive polarity, which is a brilliant pink. See that plume magnify and expand, bathing your four earthly bodies. Feel the majesty of power tempered by love. These two plumes of life and consciousness are the 'parents' of the Violet Ray of alchemy.

Step 7. Take another deep breath up and into the electric fire of the Presence and flood the yellow plume, which holds the signature of illumined wisdom. See that glorious golden light enliven your Higher Mind and inform every particle of your being. You are flooding your being and your world with true power tempered by immortal love and qualified with illumined wisdom.

Step 8. Send gratitude above to your source of life, your Presence. Like the sun that nourishes us all, the Presence pours its abundance without limit and without judgment. It is up to us how we qualify this life force and that is the nature of free will.

Step 9. Gift this energy to the Earth and feel it flood down through you and into the rich sediment and soil, to the magma and finally to the molten core, and then up and through in all directions, filling a force field 10,000 feet above, below and around the Earth, embracing *all* life upon her. Our beloved bodies who do our every bidding to the best of their abilities are comprised of the elements of Earth. Every animal, plant, mineral and elemental being are kindred and deserve our respect, love and gratitude.

Step 10. Call forth the anchoring of the Higher Mind for all humanity through their I AM Presence. Ask the I AM Presence to charge the Threefold Flame and four earthly bodies of all humanity with true power, immortal love and illumination. Giving back to life is essential.

Step 11. Stay standing or sit comfortably in a chair if you need to. Have your legs uncrossed and your feet planted firmly on the floor. Keep breathing slow, even breaths, breathing into your heart. Just concentrate on the air moving in and out. In many cultures, breath is considered Spirit and to breathe in is to *inspire*. Breath is filled with energy, also called prana or chi. Energy is the intelligent, electrical, light substance that fills the Universe, waiting to be moulded by mind, intention, and love. Now, breathe in for four counts, hold for four counts, breathe out for four counts, hold for four counts. Over time, build to eight. You can play with qualities here. Breathe in love, hold it, breathe out love, hold it. Or breathe in love, breathe out stress, dissolving it in the Violet Flame, and so on.

Step 12. Now, begin to move the breath through your body, using your capacity for focus. Feel the power of the breath enlivening all aspects of you informed by the qualities of your Threefold Flame. One day this flame will grow to such proportions that manifestation will be as easy as buttering toast. In time our ability to see the colors of energy will be restored and we will be able to read these processes and our surroundings visually. *Color is visible energy. Sound is audible energy.* You can breathe in and out of each chakra, bathing each one with the plumes of the Threefold Flame. Play with the breath. See what part of you calls for more attention.

Step 13. Again, breathe long, deep, now very enlivened breaths all the way into your root chakra. Feel the muscles relax and drop. Send the breath down your legs into the core of the Earth, fueling her and anchoring you. Breathe it up again to your heart. Now breathe up to your Crown and to the Sun of your Presence. Breathe back and forth this way, very long slow breaths. Feel the strong, loving, sentient energy touch and enliven each of your energy centers and visualize them lighting up as you do. Now come to rest in your heart. Listen. Is there a message for you? Can you connect to the 'you' that is observing your thoughts? That 'you' is your Higher Mind. Let thoughts come and go. Just gently witness.

During the day you can do a shortened version of this. Just take a few long breaths up to the Sun of your Presence and down to the core of Earth, back and forth and then settle your attention in your heart center where the radiant Presence is anchored as the Threefold Flame and rest there. It is immediately aligning and can be done in under a minute. In times of stress it is a great rescue remedy. It draws down the electricity of the Sun and the electricity of your Presence and draws up the electricity of the Earth enlivening and aligning you.

When to do it

Try to do the full meditation at least once a week. Daily is optimal. It is addictive in a very positive way. I find I need to drink from the cup of the Presence consciously at least once a day. More and more you will feel a flood of bliss and relaxation. It's so worth it. The more you do this the more your force field builds.

Exercise 9: Building the Inner Sanctuary

Time: 20 minutes.

Position: Lying down or sitting.

Supplies: For this exercise you will need pen and paper or journal and also have coloring pencils to hand.

Preparation: Creative Alchemy Meditation. Confirm this exercise is for your own highest good and the highest good of all. Decree *I AM my I AM Presence.*

How you do it

Step 1.   Sit comfortably with eyes closed. Keep breathing slow, even breaths.

Step 2.   Let your heart open and your mind expand. Now visualize yourself walking down a long path that is softly lit. You notice a mysterious door. It's glowing with the light behind it. It can be a door in a tree or on the path in front of you. Open a door between worlds and walk through. Don't question where you find yourself. If you wish, you can summon a landscape or allow yourself to discover one. What does your sanctuary look like?

The best technique is just don't hesitate, be spontaneous. It's not a test. Keep walking and see what you come upon. You may come upon your sanctuary already created and recognize it as a place you have traveled to in your dreams or you may design it from the ground up.

What type of landscape would be a fit sanctuary for your inner sovereign being, your majestic Presence? Is it surrounded by flowering gardens? In a forest? By a river? By a beach? In a beautiful city? What makes you feel happy?

Is it Egyptian in appearance, a landscape with crystalline pyramids covered with jewels and surrounded by stars so close you can touch them?

Is it a thatched cottage with animals roaming the gardens?

Is there a marble temple?

Is it super-modern and trendy with marble, glass, and wood and huge windows?

Is it made of flowers or amethysts or pure light?

Is it space age and something from the future floating above the land or sea?

Nobody's looking. This sanctuary is just for you. In the imaginal realms, anything is possible.

Now create a beautiful inner room with a table and comfortable chairs and a room with a healing table in it. Really take your time on this step and build it just as you wish. Enjoy this!

Step 3. Activate the sanctuary with heightened emotion. Fan up gratitude, delight, a joyous feeling. Release this emotion, seeing it flood from the infinite supply of love in your heart and burst like fireworks over your creation creating sparkles all over. Sit in your favorite luxurious chair and just appreciate your beautiful sanctuary. If you don't see the images yet, don't worry. That 'muscle' will develop.

Step 4. When you are ready to return, walk back through the door. Walk down the path back into your physical body. Stand up and shake it out. Stomp your feet on the ground a little.

Now that you have found your inner sanctuary you can return to it whenever you want. Your inner sanctuary is a place for refueling and retreat. It's where you go to connect with your I AM Presence and remember who you truly are, where you go to have conversations with the Higher Minds of others, cut the cords of unhealthy relationships, speak your mind lovingly and create healthy boundaries. It's where you can go to chill, bliss out, get inspired and receive important guidance and spiritual messages.

It is also a place you may have interactions with unexpected visitors, such as your spirit guides, angels and master teachers who can reach you when you are in the imaginal realm, a very real place that we enter in our subtle energy body, also called the etheric body, our dream body. It is of a higher frequency and vibration and is visible to those who resonate on a higher frequency. Trust that what you visualize with your inner eye is real. This develops over time. You may also experience flashes of inspiration and it's going to be fun to design.

Later you will build more than your sanctuary. You can build what you wish to manifest. Practiced shamans, adepts and students of the hermetic traditions can make visits to each other's imaginal realms. Studying with the Toltec Nagual shaman Merilyn Tunneshende, she explained how advanced shamans can bring back physical items from this realm!

Once you are practiced at arriving there, redecoration of your sanctuary is as easy as waving smoke. Your mind is the negotiator of this realm but it is the 'mind' of your heart, your Higher Mind aligned with your I AM Presence. Set your intention now, like a compass, to find your ideal inner home, a place of the richest beauty and comfort you can conjure.

The more time you attend to it, the more of its reality you will perceive. You will begin to develop dormant senses that allow you to smell and taste and touch in this realm. People even have told me of amazing experiences where the inner sanctuaries they have created manifested in the physical world.

This exercise should be done frequently. You will enjoy this inner sanctuary and you can go there to meditate. Again, as it becomes familiar you can just pop in but the more time you give it, the stronger your visionary capacity becomes. It is an important precursor to conscious manifestation and how we imprint the unified field.

When you are done, get out your journal or a pad so you can write down or even draw what you created or discovered. Please don't ever worry about your drawing abilities. That's not the point. Symbols are the language of the body and of the Universe. If you'd rather just describe it in your mind and hold it there, that's okay, too.

Exercise 10: Inviting Your I AM Presence

Time: 20 minutes or as long as you wish. Once established, brief visits are possible.

Position: Seated.

Supplies: Pen and paper or journal for afterwards.

Preparation: Creative Alchemy Meditation, and then close your eyes and settle into your heart. Confirm your thoughts, words and actions are for your own highest good and the highest good of all. Decree *I AM my I AM Presence.*

How to do it

Step 1. Visualize following the path to the door leading to your inner sanctuary. Enter the gathering room. Is it grand or cosy? Make sure there is a table with chairs and also a comfortable seating area. Spend some time looking around your sanctuary. If it's in nature, are there birds outside? What kinds of flowers? What are their scents like? Are there animals outside? Is it futuristic with vehicles moving above the ground? Is it a crystalline structure in another dimension among the stars? Reorient yourself and make any changes you wish. Take a seat.

Step 2. Place a chair directly across from where you have seated yourself.

Step 3. Connect again to your I AM Presence where it is anchored in your heart. Now, visualize this aspect of yourself, your inner Godself, your superhero, sitting in the chair across from you. This part of you is exquisite. The blinding light settles down and you see your true self. Feel the voltage. Look at your glorious beauty. See that you are all you aspire to be. Look into the luminous eyes of your Presence. Fan the feeling of love in your heart for your own life force. Receive the love pouring back to you. This part of you looks on you with total compassion. You're here on Earth on the frontlines of limitation, bravely forging a path back to your true nature of limitlessness and awakening from a very long sleep. Fan the feeling of gratitude. It is said the Egyptian pharaohs witnessed their own Presences and mistook them for gods. Take in its exquisite majesty. Stay here for as long as you want while doing your best to open your heart to really feel the infinite love that your Presence has for you. When you're ready ask your beloved Presence if there are any messages. Ask any questions you might have. Take a moment to write down anything you need to.

Step 4. Go now with your Presence to your healing room. Remember, it can be anything you like. The walls can be lined with radiant crystals whose songs are audible or very modern and trendy, whatever you wish. Light can be pouring in. The healing table is warm and conforms to your body. Lie down on it. Call in any other guides who may wish to assist.

Step 5. Let your Presence scan you with hands of light. Are there areas that need clearing or attention? Ask your Presence to clear you with its electric hands of light.

You can ask your Presence anything from the basics of what foods you should eat to what is your highest purpose and what would give you the most joy and anything in between. You can ask your Presence for your sacred name. Trust what you receive.

Stay in your healing space as long as you like and receive the bath of healing. When you are ready to go, give thanks. See your Presence face you with open arms. Walk into the embrace of your Presence and merge. You might feel a tingle or even an electric jolt.

Step. 6. Before you open your eyes, put your hands on your heart and give deep thanks to the universal Source. Then, stand up and gently shake out your body. Stamp your feet.

Step 7. Journal what has come to you.

Hearing My Sacred Name

Years ago, in deep congress with my Presence, I clearly heard my sacred name.

I had long stopped worrying if it was just my imagination or real, as we mostly all do in the beginning. Instead, I treasured the name and began to write a story of my life from the beginning of time, dictated from my Presence.

At a symposium in America where several of us were invited to speak about the next evolutionary step for humanity I met a woman who came over to me and asked if a certain name meant anything to me. When she spoke the name, I said it did, that it was my sacred name. She said she had heard it one day meditating on the mountaintop where she lived and also that she had received a message we would meet. This indicated to us both the depth of the connection we would go on to have.

Later, my friend decided to name a future child a variation of that name. Having lost touch with her because we lived in different countries, one day out of the blue I felt a strong nudge to call her. It so happened she was right then giving birth at her home with her partner. She had no idea why she answered the call while giving birth to the child who would share a variation of my sacred name, but she had felt compelled. A spiritual niece perhaps? Life is wondrously mysterious.

Trust the messages you receive from your Presence when you are in the sanctuary. More and more frequently life will confirm them, sometimes in the craziest ways. On future visits, ask your Presence for stories relevant to the path you are forging and write them down.

More than anything, I want you to know how magnificent and unique we all are, *you are*. Each life story is precious. It will all make sense in the end. Each of you is an exquisite rose in a great garden, becoming and unfolding. There cannot be anyone who is not magnificent, unique, needed and loved, we have just forgotten who we are. While we are unique, we are also completely interconnected, animated with the same universal Source, permeated with universal consciousness, sharing the same air and the same atoms, made of the same stardust. Waves upon the eternal sea.

> *I celebrate myself, and sing myself,*
> *And what I assume you shall assume,*
> *For every atom belonging to me as good belongs to you.* [168]
> – Walt Whitman, 'Song of Myself'

Grounding into Everyday Life

After you have practiced this for a few days, start to build the bridge between realms so it is simple to go from 'day to day' to the imaginal realm and let the imaginal begin to whisper its blessings and flow over that bridge to you. Your inner guidance is whispering to you all the time. Take a day off to practice listening to its voice. We override this voice continually. It says, 'Take a break, go to the park.' We say, *no, just a few more pages*. It says, 'Drink some water, make a fresh juice.' We ignore it, becoming dehydrated. We are receiving guidance when the whispers say, *turn right, call that person, read that book, time to change careers, make that leap, follow your soul, dare to dream, don't go there, do go here*. We ignore it. Its voice gets softer and softer. Finally, we can't hear it any more. So, taking a day to listen and act on any message is very good practice. When the connection really starts to kick in, it will give you practical messages as well as wisdom messages and you will notice they are accompanied by a little electric charge. You can train yourself to listen, but you need to give yourself time.

I know it's the voice of my Higher Mind because I get a warm feeling, or a sense of a cloak of silence descending, and sometimes a high degree of alertness. If the inner voice makes you feel anxious it's more than likely part of your patterns, what could be called the ego. With practice, it becomes easy to differentiate and then finally you hear the true voice of the Presence most of the time.

Exercise 11: Your First Guests

Time: 20 minutes, but give yourself as long as it takes.

Position: Seated.

Supplies: Pen and paper or journal.

Preparation: Creative Alchemy Meditation and then close your eyes and settle into your heart. Confirm this endeavor is for your own highest good and the highest good of all. Decree *I AM my I AM Presence. I AM the only Presence acting here.*

How to do it

Step 1. Visualize following the path to and through the door leading to your inner sanctuary. Enter the gathering room in your inner sanctuary. We are using the same procedure that you did with your Presence. You can invite anyone – an ancestor, a wisdom guide, or someone with whom you are having problems. Before you begin speaking, acknowledge the other person's Presence and speak from your own.

    Fan the feelings in your heart of love and gratitude and excitement. See the atmosphere sparkle with the electrical life force of these heightened emotions.

Step 2. Ask the questions you wish to ask. Have your pen and paper ready. What wisdom do they have for you? Do your guides or an ancestor have the answers you seek? What advice and comfort can you give your younger self?

Step 3. If it is someone you are having problems with, move to a more formal setting, such as a boardroom table. Invite their guides and yours. Speak to their Presence. Speak truthfully without blame and suggest and listen to suggestions for healing. If it is a co-dependent, unhealthy relationship you can ask your I AM Presence to cut unhealthy cords between you. I always ask the Archangel Michael to do this for me. I am no longer surprised as I once was when changes occur in others after this ritual. You don't need to cut people out of your life, you are just cutting away unhealthy attachments. Give thanks for any learning experiences, bless and release. If you need to, go into your healing room and lie on the table and call your Presence and guardians to heal you.

Step 4. Open your eyes, stand, ground yourself with some deep breaths and sit down to journal.

My Stepmother

I never had a good relationship with my stepmother. She seemed to actively dislike me. She was high society and I was a wild child. I went to live with my stepmother and father when I was 12 and we seemed as opposed in temperament as an Atlantean queen and a priestess of the grove. Perhaps there is some truth to that perception! I did not respect her values and she did not respect mine. However, I was very hurt by her rejection of me and by her efforts to humiliate me.

Over the years I did learn a lot from her. She was a great lover of the arts and knew a lot about art and theater. I learned to dress in a slightly more civilized fashion. However, she could still reduce me to tears within an hour of arriving for a visit.

During my period of radical awakening and 'divine intervention' in the mid-1990s, I began to have meetings with her in my inner sanctuary. I asked for the Presence to heal our relationship. Cords were cut and I began to see her and understand her differently. Underneath her bullying, powerful exterior, she was vulnerable and afraid. There was a very real change in our outer relationship and for the last 20 years of her life we had a loving and sincere friendship.

> *Time and space are but physiological colors*
> *which the eye makes, but the soul is light,*
> *where it is, is day; where it was, is night …* [169]
> – Ralph Waldo Emerson

Whose life are you living? Have you followed your heart? This is a call to action! When we're little, our connection to the universe is still strong. We are able to communicate with everything. We're told we are daydreaming or playing with imaginary beings, but we're not. Then we stop daydreaming. How many times did your inner kid want to dance or play and you were told not to? It's time to dance again.

Active imagination is a term that was first utilized by Carl Jung for a process of relaxed engagement with the imagination that was crucial to his technique of psychotherapy.[170] With active imagining, you are allowing the part of you that you are unaware of to come forward through the process of authentic art including painting, drawing, writing, dancing or sculpting and other expressive techniques which give form to the imagination. Jung then interpreted these images, but for our uses we will not interpret so much as allow. We will engage in the free expression utilized in Phase One, but rather than call forth an emotion we will call forth our imagination.

The purpose of this practice is to get your imagination free flowing. It helps to take your default-network mode (that hypercritical hub in the brain) offline, so our consciousness can expand and we can draw in some treasures from the cosmic storehouse. You could call this activating the Muse.

Exercise 12: Active Imagining

Time: 20 minutes to an hour.

Supplies: For this exercise, you will need poster paints, brushes and sheets of paper, pen and journal. Music is important as well.

Preparation: Center yourself and take some long deep breaths. Now put on your favorite music and dance your heart out for a song or two.

How to do it

Step 1. Assemble your pages and paints and journal. You can either engage in painting or writing. This is a very different exercise from those in Phase One.

Step 2. If you are using paints, close your eyes, visualize or feel an image and start to paint it. Don't analyze, just paint. You can also draw. Have fun. Let your imagination out of its box and play! Playing is the secret to free-flowing imagination and these techniques are increasingly being used by business to problem solve and brainstorm. More about that in Part Three of Creative Alchemy.

Step 3. If you're using writing for this exercise, just start to free flow. Let any crazy thought come. Write gibberish even. Perhaps this is how Charles Dodgson, aka Lewis Carroll, wrote *Alice in Wonderland,* very likely how Allen Ginsberg wrote *Howl* and how Walt Whitman wrote *Leaves of Grass*.

Although you may find some jewels, the real purpose is to turn off the policeman in your head so real flow can occur. That little watcher gets very confused if you're not listening to the inner critic and if you are free flowing there is nothing to criticize. Stay within the lines? There are no lines! Have fun! Play with the colors. See what happens when you drop watery yellow into blue or blue into red.

Use this exercise whenever you are stuck and whenever you are about to have a brainstorm. Definitely use it before you utilize the next exercise.

Exercise 13: Beginning the Blueprint

This exercise is a precursor to our Templates for Manifestation in Phase Four.

Time: One hour.

Position: Seated.

Supplies: Pen and colors and paper or journal.

Preparation: Creative Alchemy Meditation first. Then, close your eyes and settle into your heart. Confirm you act for your own highest good and the highest good of all. Decree *I AM my I AM Presence.*

How to do it

Step 1. Visualize following the path and walking through the door leading to your inner sanctuary. Enter the gathering room. Sit at the boardroom table. Connect deeply with your Presence. Bring to mind something you've been hoping to manifest, a dream, a project. Now, invite everyone who is going to help you with your project anywhere in the universe, connecting through the Higher Mind and Presence. If it's a film, invite Alfred Hitchcock if you choose, or Ridley Scott or Ava DuVernay. If it's finance, perhaps Warren Buffett or someone else you admire. If it's a new home invite the finest architects and builders. If it's art or sport or business or any other field, same idea. Invite those who have excelled in a way you admire. Now, expand your room ad infinitum and invite all the future readers, viewers, clients, attendees, buyers, subscribers, stadium-style. Also, invite everyone you are already working with, those you wish to work with and all the guides for the project.

Before I start any project, I appoint two divine personages who I admire and love. Since I have started to do this I have noticed how smoothly the projects go. And of course, dedicate the project to your own highest good and the highest good of all. Give deep thanks to your Presence for the life force it will supply in all manner of forms to complete this project successfully.

You are going to create the perfect blueprint for the project. You're going to pay attention to the tiniest details and conceive of the project brick by brick. Request everyone you have invited to connect with their own Presence. Together bless the project. Take the floor and explain why you want to manifest this and how it will be of benefit.

Really take your time. You may need to meet several times and you will continue to meet all through the project. Really listen when you ask a question and be ready to write down the answer. Detail is the secret here. Really take your time. Remember the architectural blueprint analogy. If one line is out of place …

Step 2. When you feel that you have gone as far as you can for this session, end by visualizing the completed project, even if all the steps still need completing. Transport yourself and the group to the site of the project's completion. Travel there in the imaginal realm outside of time and space. Look at what you have created with immense pride. Feel the joy, the sense of accomplishment, the relief, the wonder at your amazing manifestation. Flood it with your passion, excitement, gratitude, heightened expectation, the joy you would feel if this dream had already manifested. Sustain it as long as you can. Visualize the project's 'skeleton,' its sacred geometry, lit up with sparkling electricity and the project itself glowing with light.

Step 3. Create a symbol or mandala for your manifestation. Place this symbol in your heart at the center of the Threefold Flame. You can summon this symbol and bathe it in heightened emotion at various times throughout the day and, especially powerful, before you go to sleep.

Step 4. Close the meeting. Thank everyone who came, the guides and the Presence. Take as long as you wish to relax in the glory of what you are achieving. When you're ready, leave your sanctuary. Find the glowing door, go through it and walk the path back to where you are sitting in the chair.

You have formulated your dream, envisioned it, in and flooded it with emotion. You've made a powerful symbol to represent its energetic signature. It could end up being a logo and it would be a very powerful logo because the power, story, vision and outcome of your dream would be encapsulated in it.

Step 5. Get up and stretch. Shake yourself out. Make some detailed notes in your journal. Take your time. The more highly detailed the better. You can use images that your draw or cut out from magazines or find on the net. Draw your symbol carefully and put it somewhere only you can see it.

In 2017 at the Quantum Healing Retreat and Business Success Bootcamp run by Hong Curley and Mike Curley in Australia, at which I taught Creative Alchemy master classes, I was asked to create a special workshop for young entrepreneurs. I assigned them an exercise to focus and distil everything they dreamed of for their business and their future into one symbol. They all made wonderful mandalas and symbolic shapes that encapsulated their visions and then got up and shared them. It was moving and very inspirational as their true essence was made clear by the process. They were each so unique and powerful and passionate and interesting. Read on for a lovely everyday miracle.

A Visionary Symbol of Your Dream

Alex is a young man who was part of the group I was guiding in this exercise. Going deep within to the imaginal realms, he emerged with a complicated symbolic image that consolidated his vision of what he was currently working towards manifesting, a water purification system. This image from his Presence led to a leap forward with a massive inspiration that solved some problems and furthered his already innovative invention.

The next day Alex was awarded an Aspire Foundation development award worth $50,000 to develop his invention.

I sometimes see my visionary symbol emanating from my heart, multiplied thousands of times like golden bubbles to find and kiss all my finest collaborators and outcomes. I also say a prayer frequently that with love and ease and grace, please take away what no longer serves and bring to me what is needed. Be careful about that prayer and make yourself ready. It's very powerful.

Using the Manifestation Process to Reverse Fate

When my husband died I was given the seemingly impossible task of saving the theater and pub he had founded and that we had run together for 21 years. There were so many reasons to believe it was impossible. There was an enormous debt. The fabric of the 1860 Victorian building was in dire need of repair. There was a deep malaise experienced by everyone there after my husband's death. I was in terrible shape emotionally. We didn't even have funds to produce shows.

It's so important to never pay attention to the outer reality (except for obviously the day-to-day things that must be done). The power is in the inner reality. Using the story of Joan of Arc – the simple 17-year-old country maiden, who against all odds saved France from English occupation and crowned the rightful king, faithfully following the guidance she received – I gathered strength to forge forward and created an architectural template of the building in perfect shape, with wonderful, happy staff all getting along, and resources for shows and the renovation of the building. I held that vision in my mind as the outer reality lurched from crisis to crisis.

At one point, I felt physically in my shoulders that the theater wanted to grow into a larger building. I added that to the vision. Especially at night, lying in bed, I 'watered' the vision with strong positive emotion and electricity and watched it shimmer with light in my mind's eye, often through floods of tears.

It wasn't easy. But the story ended very well. Everything came to pass. The theater survived against all odds. In fact, even though I now live on the other side of the world, that template is still generating. The theater will soon grow to twice its former size and have a studio theater, my late husband's dream.

Others have reported positive results faithfully using these techniques, including physical healings, the manifestation of very big projects and success in business. Sometimes people have manifested their sanctuary!

Some also report a change in the direction. Connecting with your Presence on a continual basis you may find long-held dreams fall away. This can be disconcerting at first, but what is happening is you are aligning with your highest gifts and talents. Then there's the joy when you find what a fabulous fit the new direction is and how perfect it is for you. You may have a PhD in architecture but your heart and soul are revealing a destiny as a designer of landscape gardens. You secretly thrill at the idea but shudder as well. It's going to take courage, but you will find true happiness if you follow your heart.

Remember, your Presence is connected to everyone else's. The Presence of humanity works together to create the perfect pattern of manifestation when called to action. If we are all walking the path of our true purpose, there is no competition. Every individualized focus of life has the same end in mind, to manifest the perfect pattern of life eventuating in a united orchestra playing the symphony of a new world.

The next exercise will be accessing the multidimensional non-time-bound aspect of yourself that has already developed the qualities you are hoping to bring into your present reality but you will add some crucial steps.

Before we begin, a thought. The Second Law of Thermodynamics, also known as Time's Arrow, is the only physical law which points time forwards in our Universe. Time moving forward, future bound, is characterized by increasing entropy. Entropy is decline into disorder. If time moved backwards entropy would reverse, nothing could break, we would get younger, systems would become more orderly. Scientists are exploring the idea that time and therefore causation could run backwards in time as well as forwards. Effects could exist before causes, a thought phenomenon called retro-causality. The future would affect the present.[171] Electrons *have* exhibited non-time-bound behavior. [172] Theoretical physicist Carlo Rovelli in *The Order of Time* maintains that time passes at different speeds depending on where we are and the speed we are moving. On the most fundamental level there is no space–time and there is no difference between past and future.[173]

The quantum physicists are confirming what the sages and mystics have always told us: everything is energy. Beneath the surface of what seems solid is energy. As frequencies quicken the laws of space and time dissolve. When we are one with the broader consciousness through the faculties of intention and attention, what could be possible? If we continually use the technique of Future Memory, we could reverse entropy. We could become younger. Effects could influence causation.

Time is flexible. It can become our friend and do us favors, stretching, slowing, folding and looping.

Exercise 14: Future Memory [174]

Special thanks to Dr Norma Milanovich for this exercise.

Time: 20 minutes to an hour. As long as you need. Consistent commitment bears richer fruit than sporadic.

Supplies: Pen and paper or journal.

Preparation: Creative Alchemy Meditation. Confirm your manifestation is for your own highest good and the highest good of all. Decree *I AM my I AM Presence.*

Position: Sitting or lying. Just be very comfortable.

How you do it

Step 1. Follow your path down to your sanctuary. Make yourself comfortable. Invite your beloved Presence to sit across from you. Allow yourself to melt into your Presence where our superpowers are held. Immerse in the universal supermind able to bend time and space using the flying carpet of your imagination. Feel your electromagnetic field shift increasing your awareness of infinite possibility. Take a step into your Presence. Feel the jolt of electricity as you become one. Feel the infinite love and the coursing power.

Send love to every aspect of your being in all timelines and dimensions and at all ages of the life. I often wonder if moments of reassurance I felt as a child were healing energies I sent back down the timeline as an adult.

Step 2. Visualize yourself in a future time, united with your Presence and shining like a sun. Invite that future self to give you guidance. You can, of course, also invite your younger self at any age and pre-birth as well.

Visualize your future self as your highest aspirations for yourself realized in all areas. You did start going to the gym after all. You wrote those novels or started that business or had retrained in a different field or parachuted from a plane, whatever your dream is. See yourself in the future as calm, prosperous, joyful and in your mastery. The Presence blazes around you. You are lit up like a sun, nourishing everything around you, magnetizing everything you need.

You are very fit, healthy, wealthy, well loved and wise. You are looking back on your past few decades, for example, or your whole life. Savor what you have accomplished. See every dream fulfilled and review it in enormous detail.

Step 3. Give enormous thanks to yourself. Stay connected to your Presence. Leave your sanctuary and come back.

Step 4. Take some time to journal. You can continue the communication anytime, anywhere. Write from your heart without thinking. Give thanks for all that awaits you. Thank yourself for all the amazing work you are doing. Thank all your guides. Thank everything in your life.

When to do it

Whenever you have some spacious time uninterrupted. Ideally, once a week. When you wake up in the morning when you are still in that liminal space between worlds or relaxing in the bath or lying in a warm grassy field.

What to do

Check in with enthusiasm throughout the week. Enthusiasm is a powerful assist in this process. From the Greek *enthous,* meaning inspired or possessed by God.

What not to do

Share your process at this delicate point. That goes for all of these exercises unless you have a dedicated partner walking this path with you and you are both truly committed and ready to support each other with open hearts and no competition.

Here are some extra techniques to play with to continue to sensitize yourself and develop the muscle of your visionary imagination.

Visionary Communication

This technique assists with mindfulness and concentration and involves fully focusing your attention, learning active listening, developing confidence and trust. It involves becoming an 'other' through the process of observation.

> *The moment one gives close attention to anything, even a blade of grass becomes a mysterious, awesome, magnificent world in itself.* [175]
> – Henry Miller

You can engage in this observation with a flower or plant or tree. You probably knew how to do this as a child. Alone in a garden ideally, with plants still rooted, give yourself some time to really feel into a plant or a tree or a flower.

Get comfortably seated. Look at the plant you have chosen. Just relax and allow. Let thoughts come and go. Take your time. Soften your focus. Breathe. Let your thoughts go. Listen.

Using this technique with flowers, I entered the conscious realm of nature and communicated with plants, realizing what we call the elemental kingdom is in fact the Higher Selves of the plant realm. I was humbled by the power and wisdom of the messages I received and realized I had been arrogant. Several others learning the techniques of Creative Alchemy were also able to open up and enter into communication with plants. A tree asked my friend and colleague Christina what the sea looked like. She closed her eyes and conjured an image to show it. Others were able to intuit the remedies a plant could offer which later proved correct. I have often received profound wisdom, beauty and poetry from plants. Plants and animals seem to know who can 'hear' them and intuitively communicate with them. That butterfly that is following you may well have a message. The horse that trots from one end of the field over to you wants to engage. All of life has its own song.

> *To see a World in a Grain of Sand*
> *And a Heaven in a Wild Flower*
> *Hold Infinity in the palm of your hand*
> *And Eternity in an hour*
> – William Blake, 'Auguries of Innocence'

Please try this and know you will also be able to communicate with nature. It builds the muscle of your visionary capacity and prepares you for conscious creation. The angelic and elemental kingdom are part of the seven realms of

Earth: human, animal, mineral, plant, angelic, elemental and ascended masters. When we allow ourselves to work in concert for the highest good all of life joins in to help us. What have you got to lose by trying? Training in listening is also training in trusting. Everything alive is conscious and longs to tell its story and *everything* is alive!

Many indigenous cultures honor the consciousness of plants, animals, non-physical beings and Earth herself. After all, plants evolved 450 million years ago and some animals, such as sharks, even longer. Scientists are catching up and have now discovered that plants have brains! Not cranial brains like humans and animals, but cells that can make key decisions. These cells communicate much like our neurons and can send commands, nurture, educate, make decisions based on environmental factors and send alarms throughout their community.[176]

Trees become friends, care for their young and their old, feel pain and communicate emotions such as fear through a 'wood-wide web.'[177] Mother trees, also called hub trees, recognize their kin and send more carbon to them than others. When they are dying, they will send messages and necessary information, favoring their own seedlings.[178]

Scientific studies have revealed that it's possible dolphins and whales have intelligence equal to humans if not greater. Thankfully, most countries now acknowledge animals are sentient. Animals feel a broad range of emotion similar to ours and can innovate and problem solve. Their senses are acute and they are capable of high level perceptions.

All life is animated by electricity and as our ideas of how intelligence is communicated are challenged we may finally discover that electricity is the medium. Everything living is electrical. As beautiful as language is it can separate us from the body. As we become more telepathic as a species we may find animals have a lot to say.

When it comes time to build your vision, it is important to remember that there is consciousness in everything and if we honor and respect and are grateful for the participation of all expressions of consciousness coming together to form our vision, we may find we receive assistance from unexpected quarters. Whatever you call into form will be clothed by the elements of Earth and it's important to be aware of this and grateful for it to create the sustaining power for your vision.

Visionary Sensing

Engaging our senses is a powerful way to negotiate the imaginal realms. Get comfortable and start to imagine warm, golden oil pouring onto the top of your head. Feel the warm, silky, heavy substance pour over your crown, your forehead, your hair, your face and so on.

Become aware the oil is aromatic, you smell the beautiful scent with a deep inbreath and feel the potency of the scent travel through your entire inner body, enlivening it like a spring morning. As it flows over your eyes, imagine your inner vision gets stronger and, as it flows over your ears, warming them, imagine your inner hearing gets stronger. Feel this warm oil flow down your entire body relaxing all your muscles in your neck, shoulders, back, torso, pelvis, hips, legs, ankles and feet. Your entire body is covered in warm silky oil. Your body is relaxed, heavy, warm.

It may be helpful to physically experience this and then recall the experience over and over. When you are next in your inner sanctuary, lie down on your healing table and recreate this experience in the imaginal realm.

Visionary Flying

Close your eyes and bring a bird to mind, an eagle or a hawk – a bird that flies with ease. Keep looking at this bird in your mind's eye. Notice its wings, its body, its bright eyes. Where is it? On a branch? A mountainside? Now, place your consciousness inside the bird. Feel your wings and the streamlines of its body. Feel the life in your feathers. You're about to take off. Open your wings, feel the support of the air and glide on it. You're flying! Feel the wind in your wings! Your feathers are ruffling. Sail along on a strong breeze. What's below? What do you see?

Now imagine yourself to be a fish or an eel, an octopus or another aquatic animal. Feel the icy cold water. How deep are you? Way down where everything gimmers with phosphoresce? Or in the Bahamas in sunlit warm waters? Are you a big sea turtle lumbering at the bottom, floating up to have a look around? Feel how your shell and your soft underbelly. Now that you're on the surface, look around at the tourists gawping at you and taking photos. What do you feel?

The octopus is the oldest and most intelligent species on Earth with 10,000 more genes than humans. Feel what it's like to have three hearts and eight legs. The nervous system of the octopus is distributed throughout its entire body. Its sensorial powers must be very high. Feel the rushing water. The bottom of the sea bed.

With these steps, we are simply activating our dormant imagination, our greatest gift and worth the effort.

Summing Up – The Magical Imagination

You are learning to be a conscious creator. You are transitioning from Homo sapiens to Homo illuminatus. You are becoming one with your I AM Presence. You are a pioneer. It's going to take great focus. It's going to become a way of life. You will find your life starts to breathe you and carry you along. It gets easier, like sailing down a river. You just have to let go of the shore.

Here is a reminder of supporting techniques to include in your practice as you engage in the above exercises.

- Draw it, detail it, write down every step it will take for you to complete to physicalize your dream. Make a mental movie of it, write a story of it. As Napoleon Hill said in his book *Think and Grow Rich*, whatever the mind can conceive and believe, it can achieve!

- Build up and fan the flames of your emotion, the ones you will experience when you receive your dream, when it is manifest. Delight, relief, exuberance, intense gratitude.

- Spend time free associating. Remember everything is alive and wants to communicate including every aspect of yourself.

- If hanging out with your dream creates negative feelings, then reassess! Something's not right. Check inside your emotional bank and, if that's clear, then recheck your dream. Is it all yours? Not trying to please anyone? Show so and so 'what for'? Be loved by the multitudes? Get invited to that glamorous party? All of these are fine but can never be the reason for your dream. It has to be yours, all yours. Remember, if you've got a PhD in architecture but know in your heart you want to be a gardener, go for it! You will find sublime happiness and make beautiful (architectural!) gardens. If you're a chef and you want to be an artist, do it! If you're a taxi driver and you want to run a lovely bed and breakfast in some beautiful place, now's your chance. If you want to be a laptop entrepreneur and lie on a beach, or be a baker or a poet, you can. There is nothing too big or too small. All of these are concepts of limitations we have inherited.

- Revert to Phase One and go through those exercises whenever you feel out of alignment or if you feel any build-up of doubts, anxiety or unhealed emotion. Keeping your energy field clear, positive, uplifted and powerful is very important.

- Set up your dream state before sleep. When you are in your dream body, you have access to all imaginal realms your frequency will allow entry to. I always say, 'I AM Presence, take me to the highest places, the inner temples of learning where I will meet masters to guide my work to its highest expression and fulfilment, or where I will meet the dream bodies of my future collaborators or readers,' and give thanks. Flood all of it with love.

In Phase Three you will connect to your life force, the animating principle and the only consciousness that can act in your life, the Magic Presence.

PHASE THREE:

CONNECTION TO THE MAGIC PRESENCE

The time has come for the Son of Man to be glorified [179]
– John 12:23

In Phase Three you will learn a process to help you connect to your I AM Presence. You can call this aspect of your being your inner superhero. When this connection becomes solid, we can no longer do what we do not love. It becomes impossible. Your life becomes permeated by the truths of illumination. You become glorified. Not in the modern sense of celebrity, which is humanity's substitute for something remembered deep within, but aligned with your true majesty, sovereignty and autonomy. We succeed in making the shift of the ages.

The Son of Man referred to in the quote above is the whole of humanity. We could call it the 'Sun' of Humanity with modern-day greater sensitivity to gender. When we are in alignment we become anointed by the radiant eternal sunrise of our I AM Presence. We are restored to our true estate. This is the truth of our being not being taught to us as children. We are so much more than we have been led to believe. We have fallen into a mass hypnotism of limitation and all the problems of our world are caused by it. Here is our cure.

While your emotions supply the fuel that will make your dream a Living Vision, the Presence will provide the animating electricity. Just as a car with a full tank needs the electric spark to ignite it, this is the all-important finishing touch.

Connecting with our inner Presence not only provides the juice, it provides the direction and the process through the bridge of our Higher Mind. The Higher Mind is the stepped-down aspect of the Presence, whose lowered frequency can bridge between us 'down here' on the frontline of third/fourth-dimensional space–time and the high frequency Presence. The Presence, as a consequence of vibrating in the realms of perfection, perceives only perfection. Even in our seemingly imperfect reality the Presence permeates all life.

The more we surrender to this transpersonal part of ourselves, the easier it gets. Life becomes something we no longer strive for, but something we allow. You are that which created you. You are the Source. There is nothing that is part of the whole that can be separate from the whole.

Exercise 15: I AM Presence Activation

Time: 10–20 minutes. After a few practice sessions you'll be able to do this in five minutes.

Position: Sitting.

Supplies: None.

Preparation: Light a candle or bring home a flower, something to give this ritual a sense of sacredness. It can be very helpful to begin the Creative Alchemy Meditation.

How to do it

Step 1. Relax in a chair with your spine straight and feet uncrossed on the floor, and your hands with palms upwards on your higher thighs, or cross-legged on the floor or in the Lotus Position. The great master Swami Muktānanda tells us the lotus posture opens 72 million subtle energy channels.[180] How extraordinary we are! Personally, I have not yet mastered it, yet I can go deep in meditation. Making yourself comfortable is important. Posture is not nearly as important as intent. Take long, deep breaths, five counts in, hold for five counts, five counts out, hold for five counts. Build to nine. (Start where you feel comfortable.)

Step 2. Place your hand on your heart and breathe into it. Connect to the Threefold Flame of love, wisdom and power, the animating force of the Presence anchored there. Expand that feeling by breathing into it. Feel the warmth begin to suffuse your heart and chest. Bring to mind someone or something you love and further expand the warmth. See the brilliant pink, golden and blue plumes bathe your four earthly bodies, cleansing, healing and illuminating your entire force field. Feel into the consciousness of each plume. It will respond and expand with this attention.

Step 3. Move your attention deep within the inner chamber of your heart where the portal to the higher frequency octaves resides. Slip through the portal. Enter the eternal sea. Visualize and feel a radiant sun expanding in your heart, surrounding you. Feel its warmth. Feel its love. Decree, *Beloved I AM Presence I AM one with you,* three times out loud.

Step 4. Call for the Presence to bathe you in the sacred fires and cosmic light of its being. The sacred fire is the alchemical, transmuting, purifying, consuming fire that inspires the inner life force (also called the sacred fire and the kundalini) to rise up the spine, preparing us to unite with the Presence of all life. The cosmic light is the consecration, illumination and raising up of frequency. Light is nature's way of distributing energy. It lights the lamp in our hearts so that we may be the light of the world. Decree *I AM that I AM.*

Step 5. Send all the love in your heart to the Presence, the source and force of your life. Wait for the return circuit. Ask your Presence for the wisest masters and the highest guides. Tell your Presence you are ready to live your true purpose (if you are). Sit in receptive silence. Inspirations may come. There will be a response and a shift in your field even if as yet imperceptible. Give thanks for the life force you have drawn into your subtle energy field.

Step 6. Qualify the neutral life force with the qualities of the Presence: joy, youth, vitality, wisdom, love, abundance, health, concentration, clarity, endurance, courage, power and so on and allow this qualified life to charge you with the sentient electricity that fires your neurons, beats your heart, spins the wheels of your energy centers and flows through the river of your meridians dissolving all that is of a lower frequency and that no longer serves. Give thanks. Ask that your conviction, gratitude, trust and love be multiplied a thousand times.

Step 7. In the sacred art of ancient Egypt, India and other cultures that depict this union between the inner Godself and the human self, they are often wearing a *'coat of many colors'* or are within a *wheel of color*, a *body of rainbow light*. This is a reference to the Causal Body, the rainbow rings of subtle electrophotonic energy surrounding our I AM Presence, that you were introduced to in Chapter Five. The talents and gifts of our true essence are stored here along with the positively qualified substance of our life force from all experiences in all timelines and dimensions and all of our experiences of learning in the seven spheres or realms. When we come into alignment, we can access these stores of abundance and wisdom. Imagine a circular rainbow surrounding the sun of your Presence. In these bands of color are the potentialities for everything you will ever need. Your supply of every good thing. Decree *I AM my I AM Presence. I AM the abundance of the Presence in my hand and use it now to free all life.*

Step 8. The Law of the Circle calls for reciprocity for the balance of life. Take some time to focus and concentrate on the light pouring from the Presence and allow that light to flow outwards flooding every mind, heart, soul and body, every being and everything in your area with peace and abundance and goodness. The ascended master teachings recommend we do this for five minutes three times a day, to balance what we receive from life.[181]

Step 9. When you feel complete, let gratitude flood to your Presence, your life force, your genius and your supply of all good things. Take a few deep breaths and then open your eyes.

Step 10. Take some time to journal your thoughts and impression. How do you feel now versus how you felt before the meditation? Have you received any insights or clarifications? Ask that your Presence lays a carpet of light and synchronicity for you to walk on today.

When to do this

It's ideal to connect to your Presence 20 minutes a day, building to twice a day. It can be lovely to do first thing in the morning before the rush of the day. It is of great benefit to yourself and others to practice the Law of the Circle three times a day. As you practice holding your attention on your I AM Presence, the points of light within every cell in your body begin to respond and a deep purification and unification process begins. As with everything else, intention and feeling are paramount. However, if you can commit to consistency and do as much as you can daily and with regularity you will want and need to do this. I call it drinking from the sacred cup. It fills me up and gives me the strength and insight I need.

Again, a reminder to create a sacred space in which to build up energy. This makes it easier and draws you in quickly.

The process of enlightenment is very much a physical activity and certain chemicals are released by the pineal gland when the right and left hemispheres of our brain are in perfect balance, peace and harmony. This is what is meant by the ancient teaching, *if your eye is single your entire body will be flooded with light.* The single eye is the third eye, the activated visionary pineal gland.

Guides

We all have guides who reside in octaves whose frequencies oscillate so quickly it renders them invisible to the human eye unless they consciously step down their energy or ours quickens considerably. You will, when the time is right, be introduced to your guides internally, if you haven't been already. My story of meeting the masters within is in my memoir mentioned earlier.[182] There is a saying, *when the student is ready, the teacher appears,* and I have experienced that to be true. This can be the moment you say YES!

Take special time in meditation and ask to be introduced to the new masters who are now your inner teachers as you embark on the ultimate step of union with your I AM Presence. The moment you commit to the path of Creative Alchemy you will draw a powerful group of inner plane masters who will join or replace those already working with you, even if you are not yet aware of it.

It is helpful to write your questions after the above activation. Take a moment and without hesitating write the answers and trust the answers that come until the day when you prove the truth of the connection with a surprising answer to something you know you weren't previously aware of.

The benefit of working with the inner-plane masters is that they hold us steady and add their considerable power to ours. If they feel our manifestation truly benefits the highest good they will lend their great momentum to further sustain our activity, greatly contributing to the efficacy and perfection of our manifestation.

In writing this book and in creating and facilitating the system of Creative Alchemy over 25 years I am guided by a group of inner-plane masters of alchemy who wish for us to become ambassadors for the new Earth. They have been working for eons to uplift humanity. Much of the core information in this book came first from direct perception from these guides. It was then confirmed through intense and thorough scientific research – reverse-engineering these techniques to see why they work so well for so many. I have also gained great insights from direct experience with individuals and groups applying these techniques, and patterns have emerged that have been useful in bringing the material to a broad audience. This is a very humbling process. I have surrendered myself and committed deeply to union with the Presence and have been called to form a United League of ImagiNations of which you are now a beloved member. Together we will learn new ways of being that will alter our vision of reality dramatically and help heal our planet and our world and create exciting new ways of living.

Expect to meet the masters.

Exercise 16: Violet Fire Meditation[183]

Special thanks to the Bridge of Freedom Library for this excellent technique.

Time: 5–10 minutes.

Position: Sitting or Standing.

Supplies: None.

Preparation: None.

How to do it

Step 1. Close your eyes and take some deep breaths. Visualize a stream of violet light pouring through the atmosphere over you and your external sanctuary. Visualize the light bursting into the Violet Fire and burning throughout your home, your neighborhood, your town or city, your country and the world.

Step 2. Call for the great masters to assist you. They will answer your call.

Step 3. Decree that your hands become blazing suns of Violet Fire. 'Mighty I AM Presence, I command that my hands are bright violet suns, instruments of great healing power.'

Step 4. Stand up and sweep your body seven times from top to bottom. Gradually the tar like substance of calcified discord loosens and fades to smoke and that finally fades away as well. All your earthly bodies are able to shine with their full light and receive more of the cosmic light of the Presence.

When to do

Nightly before bed is an ideal time.

Exercise 17: Seeing the Presence in Everyone and Everything

Time: Continual.

Position: Any and all positions.

Supplies: A willing heart and an open mind.

Preparation: I AM Presence Activation, especially the first few times.

How to do it

Step 1. Close your eyes. Take some deep breaths. Put your hands on your heart. Breathe into the Threefold Flame of love, wisdom and power anchored in your heart where your Presence connects to your earthly vehicle.

Step 2. Decree *I AM the only Presence acting here. I AM all life everywhere.* Open your eyes and look around wherever you are. Sense the electronic life in all beings and things and know that it is the Presence.

Step 3. Go through your day and whatever or whoever you lay eyes on or think about, practice seeing the Presence beneath 'good or bad.'

When we can finally see the truth of the Presence of life, whether veiled by discord or radiating its true nature of harmony, we are never the same again.

When to do this

Ideally as much as possible.

More on the Science of Decree

Decree is how we negotiate the higher octaves or dimensions, that which precedes form. Decrees are similar to affirmations but infinitely more powerful. A properly stated decree is an edict, a command. Fill it with the heightened emotion of love and it will exist throughout eternity. It cannot help but become manifest in perfect accordance with our trust, focus and will. Why must we trust? Because we are powerful creator-beings projecting our life force into manifestation. What can ever be achieved if one does not believe it?

A good decree is, *I AM a being of cause alone, that cause is love, the sacred tone.*[184] When you are finally in complete alignment and merged with your I AM Presence, your cause is one with the universal accord, for the highest good of all and your own highest good. Like the eternal ellipse of *I AM that I AM*, the cause is love *and* the effect is love.

If you train yourself to perceive your being and our world as musical frequency waveforms, which in fact it is, you can begin to think along the lines of healing vibrational discord and restoring harmony in the symphony of your life. *Beloved Mighty I AM Presence, I call on the Law of Forgiveness for all ways I have harmed life and all the ways humanity has harmed life and ask the Violet Flame of transmuting mercy to dissolve cause, core, record, effect and memory of all discord I have or humanity has created in all timelines and dimensions.* Seal it with gratitude. The discordant life force comes back into neutral harmony and awaits qualification with a quality. I command the qualification of this substance with cosmic victory. I say simply *I AM cosmic victory in thought, feeling, word and action.* Any other quality will do, whatever you feel like experiencing.

Why should we do this? Aside from the obvious altruistic reason, the clearer we are of discord the more harmonious and beneficial our process of manifestation will be. We will realize our dreams and highest potential and purpose with much more ease if we are clear of accumulated discord.

You can stand in front of the sun and ask that it showers down the Violet Fire on every living being, bringing healing, relief from the pressure of discordance and the love, wisdom and true power they need. You may see a pulsing geometric pattern emit from the sun, or violet light, sometimes I see the deep, pink light of universal love emanating over all life. *I AM the Pink Ray of universal love manifested in the sun's rays blessing all life everywhere.* The Pink Ray of life is the abundance ray. Love is the magnetizing activity of life. Call for the Green Ray of Truth to dissolve all illusion and deceit and so on.

If you would like to improve your life, employing the science of decree will help you. We call into play the mighty power of our Source and supply, God-in-action individualized within us as our Presence. If your every thought was *I AM harmonious life, I AM prosperity, I AM my I AM Presence made manifest, I AM the only Presence acting here*, empowered with clear, strong emotion, conviction and the clarity of focus that holds the pattern strong and true, all the bounty you could possibly wish for would be yours. Once you have achieved inner coherence by clearing your emotions, decrees become infinitely more powerful.

Another great practice is to repeat with passion *I AM abundance.* Your results may or may not come as money. Give the universe an opportunity to improve your life in unexpected ways. Say this over and over. While you're in the shower or walking. A decree is an edict that makes a law. It has power. You are commanding the electrons to form a pattern of manifestation. You are entraining your energy field to match the frequency of the decree burning away that which is 'less than.' Bathe it with love and gratitude and the expectancy and excitement of knowing it is already yours.

For 25 years now I have stated again and again that the Presence is my banker. I feel sure of it and it has continually proved to be the case. For example, running our theater in London after my husband had died, we had taken a real dip. I refused to believe the outer reality. I knew without doubt my Presence is the provider. Out of the blue I received a call. It was the evening and someone was at his desk and was for some reason reviewing us. He asked if we could use a low interest loan. This is one of countless stories when the safety net appeared. I have noted the cosmic 'humor' in that it tends to arrive in the 11th hour. Sometimes the 11th hour, 59th minute and 59th second! Listen for the guidance, notice the synchronicities and be ready to act and stand strong in your trust.

> *If the sun and moon should ever doubt they'd immediately go out.*
> – William Blake

If you don't feel good about your body, decree *I AM fit, strong, and healthy!* You are not being false. You are imposing a new blueprint on your energy field. Stop the negative mind chatter, say the decree, and your will and intention will shift. After a while, you'll begin to walk instead of drive. You'll choose to eat an orange instead of ice cream. Your actual desire changes. Slowly, you will build yourself anew. The practical application of Creative Alchemy is very beneficial here. Comfort eating, excess drinking and apathy are all signs of suppressed uncomfortable emotions. When they are released, all those habits become Stone Age for you.

Each decree has a potent energetic signature. It takes time to entrain your energy field. Carefully choose one and use for some time before you use another. I like *I AM pure joyous life force*. I know if I AM that, life takes care of itself. *I AM the balancing breath of perfect life.*

A Visual Expression of Decree

I was shown Kirlian photos of three plates of food. This is a kind of photography that was named after the Russian inventor Semyon Kirlian, who created a camera sensitive to the electromagnetic field of life, a precursor to the enhanced Electrophotonic imaging cameras now available.

The first plate of food had no words said over it and it emanated a few centimeters of life force. A prayer of gratitude had been said over the second plate, which emanated a larger aura of life force. For the third, a decree for the food had been said. And this one had a full inch of very bright light emanating from it!

Before your meals from now on, you could say, 'I AM the cosmic victory of the life force in this food transforming into pure light energy and perfect nourishment for my body.' Enjoy finding the perfect decrees. As your energy field builds you will begin to sense their power and occasionally feel an accompanying bath of bliss.

Summing Up

The electronic body of your Presence pulses intelligent light energy. Anchored at your heart as in the Threefold Flame, it supplies the life force for your four earthly bodies. Your Presence is a mighty being of fire, the great Presence of life that dwells in a dimension of such exquisite perfection it would stagger our human intellects. As there is no impure substance at all within the light rays of the electronic body of your I AM Presence, it perfectly emits the music of your keynote. On this keynote is the energetic signature of your purpose, your gifts and your talents stored within its Causal Body. Can you see how it stands to reason that when we are clear, the magnetic attraction of that frequency can do its work, and draw to us the manifestation of those gifts and talents and purpose? With our dedicated focus, commitment, determination and action, of course.

The Threefold Flame of love, wisdom and power is the core of all manifestation. It is the nucleus and the cohesive power that draws the manifestation into being animated by the electronic energy of the Presence, filled with the life substance of our Causal Body, fueled by the power of positive emotion and designed with the architectural patterns of the Higher Mind. The sentient energy that fills the universe, the energy that Nobel Prize-winning quantum physicist Max Plank called intelligent Spirit, responds with synchronicities and opportunities.

- Know there is a destiny waiting for you. See it as a treasure hunt. Keep your heart open, be alert for guidance, stay expectant and banish doubt. Remember, life's currency is energy. The PhD of Earth university is the understanding and mastery of energy, the fabric of reality. When we truly understand this, we will be able to solve our world's problems together using *technology of mind.*

- Every morning call the pillar of shimmering white light that emanates from the I AM Presence, which protects you from discordant energy waves surrounding you with a canopy of fire nine feet in diameter. Here's a good decree for that: *I AM the cosmic victory of the pillar of light surrounding me, my home, my finance, my visions, my car, my family, my loved ones, charged with all the qualities of the Presence and flooded with the Violet Flame.* I also ask that it stops any discordant waves I might emit and zaps them with the alchemical properties of the Violet Fire. We can either be bombarded by the discordant energy around us, resulting in fatigue and other maladies or we can be masters and transmute it and redirect it. As your mental body strengthens and your power of decree builds, this shield can filter thought and emotional waves as it builds strength through daily invocation. You will find the daily use of the alchemist's ray a great assistance in cleaning up the energetic debris of life.

- At some point during the day do this for all life on Earth. Ideally three times.

- Try repeating *I AM harmony* as an inner mantra. With this you will literally alter the vibration of your cells from discordant and out of tune to sonorous and melodic. Harmony is crucial for orderly manifestation.

- Cultivate an attitude of gratitude.

- Send love to your Presence and wait for the return current.

- Walk away from the patterns of low self-worth inherent in our cultures. Accept that it is impossible for you to be less than wonderful. Know this is also true of others.

PUTTING IT ALL TOGETHER –
FOUR TEMPLATES

The Mind is its own place and can make a Heaven of Hell, a Hell of Heaven.
– John Milton, *Paradise Lost*

In Phase Four you will experience step-by-step examples of how Creative Alchemy may be applied to your dreams. Hopefully you now know that you are the author of your life, the *creator* of your reality and more importantly, truly understand how that is possible scientifically. If you have applied the first three phases, you will have begun to shift your perceptions and gained an experiential understanding of the *fluidity* of your reality. If you have applied the foundational techniques of Creative Alchemy and begun to clear the sticky, suppressed emotion and memory that can make a hell of our lives and you will have created new neural pathways that bathe your cells with beneficial chemicals and hormones and begun to access your inner guidance.

Your imaginal muscle is strengthening. You have made connection with the force of your own life, your I AM Presence, your inner superhero. You now know you are responsible for how you qualify the energy of the life force and also responsible for cleansing life force you have qualified discordantly. You have tried and tested tools for both.

You are beginning to understand conscious creation and how you can create a heaven of your life and co-create a paradise on Earth, characterized by sovereign beings aware of their autonomy and inner supply. You are finding greater harmony and also gratitude for what you already have, no matter what that looks like.

You understand Creative Alchemy is a life path, the Great Work of transforming our psyches from lead to gold and all the rewards that offers. You have joined the shift of the ages and the evolution from Homo sapiens to Homo illuminatus, and know there are habits that must now be left behind. You don't need to become a monk. Moderation is key and approaching your choices with conscious awareness.

Like countless others exploring the fruits of our current human potential movement, you want to learn how to manifest your dreams and find your true most rewarding purpose. *Whatever is in your life right now you have manifested.* You are very powerful. The good news is you are learning how to rewrite the script.

You are aware that life will continue to cooperate with your commitment to this path and offer you situations to make you aware of that which still needs to be cleared. The emotional clearing techniques are to be used faithfully whenever you experience emotional discord and this is ongoing although it will lessen considerably over time. If you are willing to commit to those simple techniques in Phase One, you will reach a state of equilibrium characterized by a spacious mind, a general feeling of safety and contentment with bursts of delight or joyfulness, compassion for yourself and others and a general feeling of optimism, hopefulness and gratitude. You will increasingly feel in strong alignment and your ability to problem-solve and innovate will have increased markedly. *This is when you are ready to begin Phase Four.*

You will have freed a great deal of the energy used to suppress emotions and should be feeling much more vitality and well-being. You will have befriended your emotions and learned new ways to experience and respect them. You will be feeling more enthusiastic.

You're learning to be braver and take risks, letting go of the shore and letting the river take you to an exceptional life, the best you that *you* can be. There's no competition in this game. Your discipline will be growing and you will be creating decrees to strengthen it further. You have strategies to eliminate those habits that contradict the new destiny you wish to create. You will be creating healthy boundaries as well because of your growing sense of true self-worth and heightened discernment, and you will respect others for their choices, whatever they may be.

Connecting to your Presence with daily meditation and practicing the visioning techniques is invaluable. You are building awareness and expanding consciousness. You need to build the muscle of your inner vision so it can withstand the report your senses are relaying from the outside world without crumbling. What I mean by that is you must make your Living Vision more real than any 'evidence' the external world offers. You may need to ignore family doubt or reports that no one is hiring in your field or that your dream is not possible or anything that may cause you to believe the odds are against you.

For this reason, don't invite disbelief and criticism by sharing your process unless it is with a colleague for your project who has walked this path with you and is truly committed to your highest good or a circle that has formed around this work. Your inner activated Living Vision needs to grow in power and strength. Like the growth of a child in the womb or a seed in the ground it doesn't happen overnight. It needs careful tending, love and continual nourishment but not too much prodding or exposure. It needs your unwavering unconditional love as much as we need the sun. You will have to be true to yourself and may have to risk disappointing others who have a different idea of who you should be and what you should do.

As long as you are in alignment on the path of your true purpose and willing to do the work your success cannot help but come. But it will come on I AM Presence time and it will come in a way that is best for your path of learning and your highest good.

Before we begin, I want to say something that may seem contradictory. You could just do this by *knowing and acceptance.* Remember the story I told where my whole skeletal structure shifted just by asking? It was a casual request. But I was in a state of surrender and trust and blissfully heightened emotion. I knew it would occur, although I had no idea it would occur in such an outstanding way. It may happen for you in a twinkling of an eye. But just as the best free-form modern dancers emerged from the rigours of ballet and the most natural actors employ reams of techniques in rehearsals, to begin with structure is a good thing. On the way to that free-flowing surrender and simplicity these exercises not only greatly assist in imprinting the field around you and shifting your own field, they further your skills. It is win-win. Like anything, the more dedication, the better results.

The shorthand is: Vision + Emotion + Presence + Action = A shift in your reality. *Thoughts do not become things without emotion.*

Don't be in a hurry. A lot of power will be released that you will be in charge of, so keep in harmony! Every quality you call in has substance, color and life force summoned from the bank of your Causal Body and also the guides in charge of the particular quality.[185] Know that the right guardians are being called and endeavor to feel into these qualities. These templates have been designed to benefit you on all levels and to create cohesion, harmony, alignment and increased frequency.

Before you begin, take some slow deep breaths into your body. Feel where tension arises. Check your mind for doubt, self-recrimination, any heaviness. If something comes up, can you recall the circumstances in which it last appeared? First appeared? Using the techniques that you are most drawn to in Phase One, the healing of suppressed emotions, begin to release these buried emotions and memories and unleash their energy.

When you feel ready, bring some flowers to your sacred space, perhaps light a candle and begin. We will start with a state of being. If we begin our new journey having anchored a state of joy, will our desires change? We often want something because we feel it will bring us Joy. What will we want if we already have Joy?

It is a good idea to make a simple phone recording of your own voice reading these steps so you don't have to refer to the text. You can pause it when you need time to visualize. Remember the quality and clarity of your visualization and the depth of your emotion are crucial. As you follow the steps to create the template for Joy you are also re-qualifying your energy field with these qualities in each step.

Template for Joy

Time: Give yourself an hour the first time. You will develop a shorthand with practice and when all the details are in place, but it's crucial to go through all steps and respect the process as you lay down the template.

Supplies: Colors, paper, pen, journal.

Preparation: A determined decree that your choice is in alignment with your own highest good and the highest good of all. This may seem like an exercise in charity but it is in fact a powerful universal manifestation code. You start cycling in the strong power of the Law of the Circle, which gives you more fuel. Decree *I AM my I AM Presence. I AM the only Presence acting here for my own highest good and the highest good of all.* It can be beneficial to do the Creative Alchemy meditation first, especially if you feel unfocused.

Position: Sitting but lying down is fine, or walking. Whatever is the most comfortable and provides the best focus. The best time is often just when you wake up still sparkling from your dream flight in the inner realms, but whenever you have a clear stretch of uninterrupted time.

How to do it

Step 1. Close your eyes. Take the journey to your beautiful inner sanctuary. Once you are there and comfortable, activate your connection to your Presence using Exercise 15. Feel the love of the Presence flood you, igniting the Threefold Flame. Your Presence loves you with a quality of love beyond our human reckoning. Send back your love to this exalted aspect of your self blessed with all the qualities of the universal Presence. Feel your energies merge as you move closer to full integration with your Presence. Feel its huge light warm you like a sun. Call in the guides and guardians who are willing to offer their momentum to assist with this project.

Step 2. Call upon the power of the *will*. Set your intention unwaveringly. Hold your goal in mind with laser focus. Your goal is to experience Joy, an essential quality of our life force many of us have buried. Your goal is to *become* Joy.

Step 3. Call upon the power of the unconditional *love*. The love that coheres and magnetizes. Feel this love. Put your hands on your heart and breathe deeply. Bring to mind an image of someone or something you love. Allow that feeling to flood your entire body. Now, replace the image with yourself. Continue to breathe into your heart fanning out the feeling of love and as you breathe out, heighten the feeling throughout your body.

Step 4. Call upon the power of *illumination*. Summon from your memory banks and your cellular memory when you last felt Joy. Recall feelings of Joy when you were little. What made you feel this quality? Write down some notes or draw how your life would look if you were in a state of Joy. Create a strong image of yourself in a state of Joy. This is very important. Take your time.

Step 5. Hand over your vision for Joy to your Higher Mind to elevate into Living Vision, a refined geometric blueprint of light, emanating sound waves of a high fine frequency and vibration. This musical masterpiece of quantum harmonics is new song. Light is the alchemical key and the core substance of all manifestation.

Feel the certain excitement and the gratitude of receiving pure limitless Joy and fuel the template with this excitement and gratitude. Visualize and feel the radiant blueprint come alive with your passion. *Give yourself permission to live a life of Joy*. Whatever has happened, whatever you have done, nothing takes away this privilege and right for all of us. If guilt or shame comes up, please stop and go back to Phase One.

Allow your Higher Mind to imprint this pattern from the infinite intelligence onto the memory field of your subtle energy body and the fields of your mental, physical and emotional bodies consuming previous patterns with the cosmic fire *that does not burn but transforms all it touches into itself.* Decree *I AM the light of the Presence that never fails! I AM pure radiant Joy!*

Step 6. Feel the expansion of the Presence in your heart filling it with warmth. Breathe into the Threefold Flame, feel it expand, its plumes bathing you in love, wisdom and power. Decree *I AM pure joyous life force.* Say it out loud three times, electrifying and animating the blueprint with the ancient science of decree that draws Source energy forth. Fill the new pattern of being with pure electronic potent life force from the Presence. Feel the power of will activate it and the power of love begin to magnetize what you need, and the power of illumined wisdom ensure you perceive all the steps you are given to realize it.

Step 7. Invite into your sanctuary anyone who can help you achieve pure radiant joy. Experts. Guides. Masters of the upper dimensions. Connect to them through their Presence and yours. Illumined masters of the elemental kingdoms of Earth, many of whom have never lost this precious quality of Joy, despite the tragedy of our violent interference into their realm.

Thank everyone deeply for being on your team. Even if you have not yet had signs and synchronicities, know they have responded to your call. Becoming all you can be serves humanity as a model, imprinting the planetary field with a new idea. Wake up from your dream of limitation! Really feel gratitude. It has a powerful effect on your energy fields and is also very appreciated by those who guide you! Listen. What do they have to say? Is there guidance you need to hear? Habits that needed to be added or deleted? Perhaps you will be given an ancient mantra or a jolt of pure light. Be open.

Step 8. Invite your future joyous self. Take yourself in. What's different? More sparkle? More energy? Fuller integration with your shining Presence? More positivity? More generosity? More light! Have a conversation. Is there any guidance or advice you need to follow? Anything you need to let go of or add to your life? What does your joyous self advise? Sit across from your new you. Look into each other's eyes. Feel the electricity of the Joy snap and crackle between you. Notice how youthful your joyous self looks and how alive. Ask this joyous version of yourself what thoughts or actions will accelerate Joy? Admiring and truly loving the image of your joyous beautiful self, release a replica of your Threefold Flame from the focus of your Presence in your heart into your new joyous self. Bathe your joyous self in the pink, gold and blue rays of love, wisdom and power. Take a quantum step and merge into your future self. Feel what it's like to be joyful. Feel the electric tingle. Feel gratitude.

Step 9. Create an image or symbol for your joy. It will contain the essence of the activated template. Take your time. Place it in the center of the Threefold Flame in your heart. Feel and sense the flames leap with acceptance. Surround it with the Violet Fire to protect it from discord.

Step 10. Visualize you and your world filled with the consequences of Joy. See the light of Joy flooding you from the sun of your Presence. How radiant you are! Feel as much as you can summon. Breathe this pattern throughout your body. Anchor the symbol you made in each of your energy centers repeating your decree, *I AM pure joyous life force* each time. The more visual clarity and feeling you can summon, the better. *I AM the Presence of joy flooding my being and my world.* You are literally rewiring your brain and re-entraining your energy field.

Step 11. Call in the power of *peace*. Seal it! Decree *The Will of the Presence is complete! I AM the harmonious sustaining action of this decree.* Trust. Let go of expectation. Know it will manifest if you follow your guidance and take the steps it provides.

Step 12. Create a Future Memory to do every day. For example, it's a few years from now. Everything you need comes to you. You are a joyous magnet for joyous circumstances and people. You don't even have to think of what your purpose is or will be. You embody it. You have connected to your essence and its wisdom pours through you. You are full of vitality and naturally uplift everyone near you. You follow your inner guidance and create projects when you feel like it. You are unburdened in every way. Abundance flows and you share it and even more comes.

You can't believe life could ever be so good. You are gratitude personified. You have a splendid party to celebrate. Write out the details or speak it into a recording device to listen to while you jog or walk or daydream. Really take your time here.

Step 13. Thank everyone, your Presence, your guides, all the qualities and their guardians, all the experts and your own future self. Staying united with your future joyous self, leave the sanctuary and return to your chair.

Step 14. Journal and draw ideas and images from your session. Make a physical representation of your symbol. Write down all the action steps you need to take.

Step 15. Spend some time with your Presence calling in the quality of joy for all beings everywhere to complete the Law of the Circle – ideally five minutes three times a day. The more you give the more you receive.

Step 16. Keep your symbol in mind surrounded by the Violet Fire. Place physical representations where you can see them along with images from any times of your life that are joyous, and other joyous images. Breathe your symbol into each of your energy centers. When you have time, breathe the entire Living Vision of the Template for Joy into each of your energy centers, perhaps one a day. If you feel a bit of resistance anywhere, go into process. You may find that wistful feelings come up to be cleared. The force of power inherent in the process will cause catharsis for buried emotion that opposes Joy.

When to do this

Once you have completed the process completely you can leave this template and engage in the essence of it, creating a version that emerges from your own Presence. You are creating a new way of being. You are altering your frequency, entraining it to a new one. It takes a while to create new neural pathways and change the frequency of your DNA and biofield. Step into your vision of your joyous self frequently, fueling it with the gratitude of having already achieved this transition. Make a habit of appreciation. We are joy at our core. It will come. Add activities to your life that are pleasurable. Make gratitude lists. Stop and notice the detail of nature. Take yourself out for special days to do silly fun things. Whatever helps lighten you while you become the quality of Joy.

What not to do

Don't tell everyone you are doing this. Let the seed grow inside you. Let people begin to say, *What are you doing? You look great, you seem so energized and happy. You never seem irritated or annoyed and you never blow up anymore. What's your secret?*

Don't override discernment and do things that will sabotage your vision. Discipline! You will have to let go of habits directly opposed to your goal, slowly but surely. We often keep knocking our head against the proverbial brick wall and wonder why we still have a headache!

Things to remember

Make your visualization as crystal clear as possible. That is key. Use physical representations such as photos and magazine cut outs to help you if necessary. Fan your emotional fuel. The excitement, passion and gratitude of receiving. Small daily practice is effective and when you have time you can immerse yourself for longer. Remember you are literally creating a new harmonic, a geometric pattern of that which you wish to create.

The interesting thing is that if you excel with this Joy Template you may not want to go to another one, although there is no right or wrong in it. As I mentioned, every material thing we want is really because we want the feelings we think it will generate, and the feeling that the majority of us want most of all is joy. Being in a state of Joy also oils the wheels of manifestation more than any other emotion. So, future templates for manifestation will be greatly assisted by anchoring this emotion.

The great masters and sages tell us this is our underlying emotion, a keynote of the essence of life. We have thoroughly trained ourselves to believe otherwise. When the great avatar Paramahansa Yogananda was physically suffering, one follower expressed immense distress that his beloved teacher was in such pain. Yogananda took his hand and transferred his core state of being. The follower's eyes lit up with amazement. Underneath the intense suffering ran the rivers of eternal 'ever-new' joy.

You have the master template for Joy and can use your intuition to add elements to the following templates. For that reason and every good reason, please do the template for Joy first. Now for a template on a subject that popular culture obsesses on. Remember, radiant health and inner light are the best cosmetics.

Template for Health, Youth and Beauty

Time: It takes an hour depending on how deeply you go with it. Please give yourself time.

Supplies: Colors, paper, pen, journal.

Preparation: Take some slow deep breaths into your body. Feel where tension arises. Check your mind for doubt on this topic. Breathe into the physical feeling and the mental thoughts.

How did your family feel about the physical processes? How do *you* feel about them? What are the stories playing as a continuous tape in your subconscious? If your treasure hunt uncovers buried emotion go back to Phase One. Using the techniques that you are most drawn to begin to release these buried emotions and unleash their energy. When you feel that you've released the suppressed emotion and memories come back.

A determined, definite choice to be your best self physically and a decree that this is for your own highest good and the highest good of all. Ground and connect yourself with the Creative Alchemy Meditation. Decree *I Am my I AM Presence.*

Position: Sitting or whatever is comfortable.

How to do it

Step 1. Sit comfortably. Close your eyes. Take a trip to your inner sanctuary. Connect deeply with your I AM Presence. It's important to go through Exercise 15, the full I AM Presence Activation, especially in the beginning. True beauty is when we are radiant. Summon the qualities of a beautiful, healthy, youthful, flexible, firm and fit body from your Presence. A good decree would be *I AM the perfected blueprint of my Body, Mind, Emotions and Energy Body made manifest now in its full, radiant, beautiful glory.* The ascended masters use the term Immaculate Concept for perfected blueprint.

Step 2. Call to your Presence to create permanent Health, Youth and Beauty. As you are calling forth this physical shift ask for the intelligence of your Presence to impress the pattern on your four earthly bodies and flood the blueprint with the alchemical key of Cosmic Light. Ask for these new vibrational patterns to consume all others. Decree *I Am the Light of the Presence that never fails.*

Step 3. From energy comes atoms and from atoms come molecules. You are calling forth a new crucible to hold your eternal spark of Life, the true pattern that has lain dormant of your highest expression of Health, Youth and Beauty. A powerful decree is *I Am the Resurrection and the Life of Perfection.* This is a powerful alchemical statement beyond all religion. The elemental kingdom uses the flame of resurrection to bring nature back to life in the Spring. We can apply this powerful decree to our health, our wealth and any part of our being and world that is faltering, with marvellous effect.

Step 4. Invite helpful friends into your inner sanctuary. Discuss this goal. You don't need to recognize their faces. The best trainers and nutritionists available, for example. Discuss your programme with them. Thank them. Remember, trust. Many Creative Alchemy students have expertly diagnosed their chronic maladies and receive precise guidance having never done this before. *You can do it too.* Create the appropriate decree to become your mantra such as *I Am Eternal Radiant Health, Youth and Beauty.*

Connect with the Presence of everyone present, where all information is encoded and available. Give thanks.

Step 5. Invite your new self to your inner sanctuary. Sit across from each other. Look into each other's eyes. Admiring and truly loving the image of your healthy, beautiful, super-fit and radiant self, release the pearl of infinite price, the Threefold Flame from the focus of Presence and anchor the source of consciousness and life in your heart into your future self. Bathe your ideal self in the pink, gold and blue Rays of love, wisdom and power. Take a quantum step and merge into your future radiant, healthy, vital, youthful self.

Step 6. Create a Future Memory to do every day. For example, it's a few years from now. You look and feel amazing. You have that flat tummy you've always dreamed of. Your posture is fantastic. You look radiant, strong. You feel great. See yourself in all situations. People respect you more, listen to you better. Your family and friends are proud of you. You've taken up sports or special physical activities as hobbies you never would have dreamed of earlier. You can play with your kids/nieces/nephews/grandkids in rough-and-tumble games.

Instead of going to the pub, you go to the gym and find there are some great people there. You make some new friends with your same goal. You're now an expert on nutrition, especially what your own body needs. You're out in nature a lot. You need a new wardrobe. Not only does your old one no longer fit, it's no longer you. You're revitalized and more alive. You want to wear clothes that express who you are.

If you are dealing with an illness, listen to all the suggestions of your Presence but please continue with your medical care. This can support and accelerate it. Listen carefully for guidance.

Step 7. Make a symbol or mandala of your healthy, youthful, beautiful self and put it in your heart. Send it love several times a day. Breathing in and out of your energy centers saying your decree, place the symbol in each one. As demonstrated by the review of our energy centers in Chapter Two, each center connects to a nerve bundle, a gland of endocrine system, specific organs and emotions. By doing this you imprint your decree on each area for healing and entraining.

Step 8. Give gratitude to your Presence and all those who came and those who are supporting you in the inner and outer realms. Gratitude incorporates the energetic signature of receiving. Remember the old saying, 'Gratitude brings Grace.' When you are ready, leave your sanctuary, come back the way you came.

Step 9. Seal it with Peace! Decree, *The Will of the Presence is complete!* Bless it with harmony to sustain it. Trust. Let go of expectation. Know it will manifest if you follow your guidance and take the steps you are guided to take. Don't lose heart. Take a certain number of steps every day. Don't try to do it all at once and then burn out. While working on your goal, practice Radical Acceptance and love your current self with all your heart. This is crucial.

Step 10. Write down all the things you need to do based on what you know and the advice you have received. Be ready to *listen* to your inner guidance and *take action* when you are guided. Be *prepared* for synchronicities. You *must* take action when guided. Also, draw your symbol and put it on the fridge or bathroom mirror or above your bed – wherever you'll see it frequently.

Step 11. Activate the Law of the Circle. Call for all living beings to experience more health, youth and beauty.

When to do this

Once the template is set, every day. It should be very enjoyable with increasing results. Repeat your decree as a mantra. A good decree is *I call upon my real self, my eternal self, the I AM Presence to come forth and produce renewed and sustained youth, health and beauty and I give thanks.*[186]

What not to do

Don't tell everyone you are doing this. Let the seed grow inside you. Let people begin to say, *What are you doing? You look great, you seem so energized and happy.* If you receive a compliment just smile and say internally, *Thank you, Mighty I AM Presence.*[187]

Don't override discernment or you will do things that will sabotage your vision. *Most importantly you must stop all self-criticism.* This is an indulgence that will absolutely cancel all your work in the energetic realm before it has any chance to impact the physical. Discipline! Keep an eye on all of the habits that oppose your goal and refer to exercises in Phase One if you find you are self-sabotaging. Remember, slowly but surely. Incremental progress is best. Slow progress builds a mighty foundation. But timing is dependent on your commitment and your true destiny held in your perfected template by your Presence.

Here is a sample template for Prosperity and Abundance. You will see they vary only slightly. Once you are in the swing of it, your I AM Presence may start giving you suggestions to alter the template to call in other things. Trust this. Also, choose only one template to do for some time until you see results. Otherwise it's like putting in an order for several meals at a restaurant.

Template for Prosperity and Abundance

Time: Initially about an hour. Then, 20 minutes to an hour.

Supplies: Colors, paper, pen, journal.

Preparation: Take some slow deep breaths into your body. Feel where tension arises. Check your mind for doubt on this topic. Breathe into the physical feeling and the mental thoughts.

How does your family feel about wealth? What imprints do you carry? If your treasure hunt uncovers buried emotion go back to Phase One. Using the techniques that you are most drawn to begin to release these buried emotions and unleash their energy. When you feel that you've loosened and released the suppressed emotion, memories and imprints come back. Correcting 'poverty consciousness' is a crucial step in empowerment.

Make a determined, definite choice to be your best self, the top of your game in the areas of prosperity and abundance. Confirm your goal is for your own highest good and the highest good of all. You are going to inspire so many people and be a great role model. Ideally, anchor and connect with the Creative Alchemy Meditation. Decree *I Am my I AM Presence.*

Position: Sitting.

How to do it

Step 1. Sit comfortably. Close your eyes. Follow your inner path to your beautiful inner sanctuary. Hold your goal in mind – to call more prosperity and abundance into your life to enhance your life, help fulfil your dreams and enhance the lives of others. Prosperity generally means financial wealth, but it can mean being in the flow of all good things. Abundance means all areas of your life are full. When we are full we can be generous without running into deficit. If we call for abundance, we allow the presence to surprise us. What we need may come in unexpected ways instead of receiving the money first.

Step 2. Connect to your source of life, your I AM Presence.

Step 3. Call to your Presence to create and release a blueprint of Prosperity and Abundance. Ask for the intelligence of your Presence to impress the pattern on your four earthly bodies and flood the blueprint with the alchemical key of Cosmic Light. Ask for these new vibrational patterns to consume all others.

Step 4. Call for the Cosmic Light, the life force, the electronic fire of your Presence to weave the blueprint into form with atomic lace. Decree

I Am the Light of the Presence that never fails.

From energy comes atoms and from atoms come molecules. You are calling forth a new crucible to hold your eternal spark of Life. A powerful decree is *I Am the Resurrection and the Life of perfection with regards to my finance and abundance.* This is a powerful alchemical statement beyond all religion. You are decreeing for the resurrection of your birthright of opulence in your hand and to use it now for the highest good of all. You have decided to be part of the solution and not part of the problem. You will be able to be generous, activate projects and live like nature reflects for us, with infinite abundance.

If you feel any tugs, please go back to Phase One and do some work. We have had poverty consciousness impressed upon us for millennia. Get used to scanning your body and develop the sensitivity to know where the feeling is. If it's your root energy center, common for money and survival insecurity, go back to the energy center chart to help target the emotions that need to be healed and then apply your favorite technique from Phase One. Is it in your sacral energy center? You could be afraid of your own creativity due to a childhood incident. Our creative power is necessary for all manifestation. We need our creativity to live a life of abundance. Keep going on the treasure hunt until you identify and heal the source.

Step 5. Connect with the electronic signature of every dollar/yen/rupee, etc. of your future wealth, if that is what you are going for. It's good to have a specific figure. Send love to each one. Confirm you will be a wise and prudent millionaire (or billionaire), a good guardian to each unit of prosperity and that you will be very generous. Remember the importance of tithing and good works in the building of wealth. Remember as well that abundance can come in the form of what you wish, and not just a pile of money. Money is the intermediary, a neutral energy waiting to be utilized.

Step 6. Create the appropriate mantra for this manifestation. This could be *I Am Infinite Wealth, Prosperity and Abundance* or *I Am ten times ten what I need in my hand and use now.* You must say it continually so that it becomes your inner mind tape gradually deleting all negative doubts and fears. Keep utilizing *I Am the Resurrection and the Life of perfection for my finance and abundance.*

Step 7. Invite helpful friends to your inner sanctuary. You don't need to recognize their faces. The best advisors and strategists available on the planet for example. Who do you really admire? Who is doing great things with their wealth? Discuss your programme with them. Invite your future partners if you will need partners. Decree your intention to work with people who have advanced skills and are very honorable. Be very clear about that. I once manifested large amounts of money for a project using decree and the Creative Alchemy principles without that protection and it wasn't good! Connect with the Presence of everyone present, where all information is encoded and available.

Invite your future abundant prosperous self. What seems different? More confidence? Beautiful clothes? Healthier? More radiant? More relaxed? Happier? Admiring and truly loving the image of your abundant self, release the pearl of infinite price, the Threefold Flame from the focus of Presence anchored in your heart. Bathe your ideal self in the pink, gold and blue Rays of Love, Wisdom and Power from your heart and then place a replica of the Flame in the heart of your future self, imbuing it with life. Take a quantum step and merge into that future radiant healthy vital youthful self.

Make a symbol or mandala of your healthy beautiful self and put it in your heart. Later you can breathe this symbol into each of your energy centers for healing and entraining.

Step 8. Create a Future Memory to do every day. It's a few years from now. You are confident, successful and the master of your destiny. See yourself in a field or a huge, beautiful meeting room. It can be your inner sanctuary grown large for this event.

You've invited all the people whose lives you have affected positively with your wealth and the innovations it has afforded you – all your colleagues, even ones you will never meet or know who have oiled the wheels of your prosperity. Perhaps you have a charity or a foundation. Maybe you've built your dream house or made a feature film or traveled around the world.

See all your friends and family who have been inspired by you and all the people you will meet socially as you travel to exotic places. Your abundance is so great you now donate to charities the world over. See all the beneficiaries here as well. You can afford to live completely ecologically and perhaps you have created some stellar innovations to encourage holistic and ecological living. You're tithing 20 percent of your income, you have so much. You've got time on your hands and you've taken up an art or another hobby. Perhaps your paintings are on the wall or you pick up an instrument and play for everyone. You are super fit and have a personal trainer and take your holidays in the beauty spots of the world.

Maybe you burst into song in the perfect keynote of your essence. Everyone joins you, calling forth their own sacred keynote. Go around the room hugging everyone and shaking hands. Feel strongly the pride, the joy, the delight, the awe and the gratitude.

When you are ready give thanks to everyone and come back the way you came, still merged with your future self.

Step 9. Seal it! Decree *The Will of the Presence is all encompassing and complete!* The more you can surrender your human will the better. Bless it with peace and harmony to sustain it. Think of it as a garden. Seed it, water it, give it the sunlight of your Presence. Trust. Let go of expectation. Know it will manifest if you follow your guidance and take the steps it provides.

Step 10. Give gratitude to your Presence. Gratitude incorporates the energetic signature of receiving. *Really feel it.*

Step 11. Write down all the things you need to do based on what you know and the advice you have received. Make a physical expression of your symbol and pin it where you can see it. Send it love several times a day.

Be ready to *listen* to your inner guidance and *take action* when you are guided. Be *prepared* for synchronicities. *You must take action when guided.*

What to do

Educate yourself as much as possible. Seek external mentors. Carefully follow all the guidance you receive. Read relevant books. Keep a hold of the feeling of abundance and prosperity. Make your home look beautiful. Get rid of anything in bad repair that is beyond fixing. Make sure your clothes, right down to your underwear, are fresh and clean and in good shape. No gray knickers! Get strong and fit. Walk tall. Tithe to good causes. Tithing has a mysterious way of magnetizing abundance. The usual amount is ten percent. Think of all the good and generous things you can do with your abundance. Create a recording of this process interspersed with your chosen mantra and play it as you go to sleep.

What not to do

Don't tell everyone you are doing this. Let the seed grow inside you. Let people begin to say, *Wow, something's changed, what is it?* Don't override discernment and do things that will sabotage your vision. When you feel doubt, practice Exercise 12, Active Imagining. Keep an eye on 'poverty consciousness.' Discipline! Catch yourself when you revert to limiting stereotypes passed on by family or culture. Work through these when they come up. They are tenacious. The very best thing we can do is be part of the solution and not part of the problem. Keep using your decree as a mantra and activating your symbol in your energy centers.

You can see how this template can be tailored for each goal. If it's a book, meet with your editor and publisher in your inner sanctuary. Hold a party for your millions of future readers while doing the Future Memory exercise. See how they have benefited from your book. Perhaps the world has benefited. Why not? See the whole world improved as your book finds its way into the hands of millions of waiting readers. You've done so well you have a foundation to assist humanity in some way. You're traveling the world promoting your book. Or you're lying on a beach writing your next one. Publishers are seeking you out and offering advances. Create a strong decree that everyone who reads your book will be greatly enhanced on all levels and call to the masters and guides to enforce it and guide your readers.

Don't confuse the Universe. *Use one manifestation practice at a time until the goal is realized.*

A Salutary Tale

I had been working with the I AM Presence for ten years and many miraculous events had occurred. I had been in theater for 20 years and film for about ten years. I had been decreeing for the abundance to make my feature film for some time. I'd made a beautiful highly detailed vision board and had experienced many boons including a good response from highly noted actors, which necessitated frequent trips from London to LA, and some extraordinary creatives joining my team.

I had been working to clear myself for some time. I had forgiven the main players of my severely traumatic childhood but didn't yet really understand the concept of the inner mirror on a deep enough level. I still felt like a victim at times and didn't understand that 'dance' as deeply as I eventually would. I had not yet attained the complete truth of self-responsibility for the reality I inhabited. I therefore couldn't really understand why I often drew bullies into my life. Bullies who wanted to crush me, crush my work and take from me. Was I born under a bad star?

It wasn't until a few years later, working with a powerful Ghanaian shaman, that I had a vision that identified a suppressed aspect of my beloved father that was expressing in these people, always men. My father was 'the man with a heart of gold,' brilliant, the youngest graduate of Harvard of his day, but he was also bipolar. Bipolar is usually engendered by a split in the psyche. He was the firstborn son of immigrants who expected him to integrate the family into society and he ended up becoming a lawyer but with the heart of a poet. Way down deep, he was angry. This played out in patterns in his life that did not appear to be anger but were troublesome for those around him.

In the vision I saw how angry my father was and how this had pushed his energy up into his brilliant mind and also pressed it down into his root energy center and though he was deeply loving, there was an aspect of his heart that was veiled and armored.

I recalled how he had been my everything as a girl but also many controlling events where he used his power to shift my destiny. It was the 'droit de soigneur' that was typical of many men of his generation. These men I drew into my life expressed my father's suppressed anger! I had something they wanted and they were willing to crush me and what they coveted if they couldn't have it. Most importantly, I created the situations by giving my power to them. Like my father, they were always very bright, very mathematical with a strong male side and very controlling. I was weak in my male side and oblivious to the role I was playing by conjuring them into my life.

The decrees of abundance did work and I ended up with an extremely healthy budget for my film and some terrific actors. The film was made in 30 degrees below freezing weather in Lithuania, which was challenging. My husband was very ill and the timing was awful. The UK film industry had just collapsed and we winkled through by a hair's breadth. The circumstances could not have been more difficult. What was even more challenging was the one player who began to work against me, creating major problems. My very own Frankenstein.

I had begun the film project intent on testing my theories of alchemy. Any problem that arose I would transmute into a solution. Remember the saying, 'When two or more are gathered, I Am there,' by the great Alchemist? That is a quantum logarithm. When two or more are working with the creative alchemy of the Presence, the power is exponential. All my allies and co-creators I had lovingly developed the film with were either suddenly not available or were taken off the film until I was a lonely 'one.' There was no 'two or more.'

I did not yet fully understand how I created my reality and I hadn't completed the necessary inner work before embarking on such a big project. The film was made and distributed but bore the same split the team suffered. I am sure this was an important but very painful lesson on two counts.

Firstly, the inner discordance of unhealed emotion has the power to derail your template! I fell prey to the self-sabotage I now teach people how to avoid by inner clearing.

Secondly, for a big project we need others who hold the same vision. This is part of our evolution from Homo sapiens to Homo illuminatus. We are returning to unity consciousness with the awareness of individuation and remembering community.

I crawled back to theater licking my wounds and there regained my confidence producing, directing and adapting the most loving production of Peter Pan *ever mounted. I learned so much about alchemy with those two experiences. People young and old returned six to seven times to see our* Peter Pan. *They could feel the love! We were a team of 40 people with one goal, working from the heart. Inspiration flowed and abundance. We cannot always go it alone.*

Many of you will want a Template for a partner and many of those I work with believe a partner will cure all their ills. I am always very hesitant about this. The reason being that so often people are trying to complete themselves through another person. We have built up the idea of romantic love as the be all and end all, sometimes forgetting that love comes in many shapes and forms, all very important. The loneliness we feel is disconnection from our Presence. Make that connection strong first.

Remember, a romantic partnership is a hothouse or crucible of transformation above all. If both parties are able to see 'the mirror' and accept responsibility for their emotions, it can be an acceleration of growth and a powerful liaison with the potential of fully conscious sacred sexuality and joyous companionship on the path of the Great Work. But if you are not complete in yourself or at least working hard to be, phew, your relationship can be a harsh teacher and derail you. The great mythologist Joseph Campbell called romantic partnership one of three *hero's journeys*. He meant the heroism of ruthless honesty and self-responsibility where two people agree to not project or blame but use the intensity of romantic love as a searchlight on the remaining unhealed emotions that we know we have when we are triggered by another.

If you have reached autonomy and sovereignty, having a partner becomes a different exercise, one of sharing of unconditional love, that immortal love whose magnetic cohesion holds the planets in their orbit and walking together the path of the Great Work.

I give you this template in the hopes you call in either your perfect partner or the inner marriage where you become complete and autonomous.

Template for Calling in a Partner

Time: It takes 20 minutes to an hour depending on how deeply you go with it. There are a few parts so it can be done over a few days in two parts, especially the first time.

Supplies: Colors, paper, pen, journal.

Preparation: Take some slow deep breaths into your body. Feel where tension arises. Check your mind for doubt on this topic. Breathe into the physical feeling and the mental thoughts.

How was your parents' relationship when you were growing up? This is most likely our current template we need to shift if necessary. How did your parents treat you? Each other? Were they open and unconditionally loving? Supportive? Did they believe in you? What were the patterns between them? Do you recognize these in your own relationships? Sometimes we unconsciously imitate or attract the opposite. Did you have an absent parent? One who was abusive or overly critical? If your treasure hunt uncovers buried emotion go back to Phase One. Using the techniques that you are most drawn to begin to release these buried emotions and unleash their energy. When you feel that you've released the suppressed emotion and memories, come back.

Confirm you are autonomous and sovereign. You are looking for someone to share life with, not complete your life or fill in the holes. It can be a good exercise once you have healed any traumatic memory and emotion to marry yourself. Literally create a beautiful ceremony with marriage vows that you won't abandon or betray yourself, you will love and honor yourself and so on.

Strongly decree that this is for your own highest good and the highest good of all. Creative Alchemy Meditation. Decree *I Am my I AM Presence.*

Position: Sitting.

How to do it

Step 1. Sit comfortably. Close your eyes. Take a trip to your inner sanctuary. Hold your goal in mind, a healthy, communicative, fun, loving relationship with excellent reciprocity. Someone who will walk the path of the Great Work with you. Consciously connect to your source of life, your I AM Presence, using Exercise 15.

Step 2. Invite any previous partners to the boardroom. Communicating through the Presence, talk about the relationship. What was good? What didn't work? Give thanks for all the good and all the learning. Now, cut the cord between you. Invite your parents and do this with them as well.

Now, think about all the qualities you would like in your partner. Be very detailed. Look at the list. If you are calling in someone very fit and healthy, are you fit and healthy? If you see this person as a stylish, sophisticated person, are you that way? Are you living up to your highest goal of yourself? That must come first. Come back when you are ready. Get your journal and write notes. Write down all the qualities you would like your partner to have in detail.

Step 4. Go through your house. Are there lots of ornaments sitting on their own or paintings of solitary images? These all need to be changed. Pair ornaments and make sure you have images and statues of loving couples. Everywhere you look should re-imprint your psyche with the idea of a love relationship. Move out half your wardrobe so there is room for another person's clothes. Getting rid of clothes that you don't love is a very good thing to do anyway. Make sure your bed is big enough for two.

Step 5. Create the appropriate decree for this manifestation. These could be *I Am in a loving relationship. I Am sharing my life with the partner of my dreams. I Am worthy of deep love.* Say it continually so that it becomes your inner mind tape, gradually deleting all negative doubts and fears.

Step 6. Conjure up some really positive emotions. Gratitude for the love you may already have in your life is important.

Step 7. Go back into your inner sanctuary. Ask your Presence to create the template for the perfect relationship for you right now. Feel and visualize the template forming. Take some time to make a list of all the qualities you wish your partner to have.

Step 8. Invite your new partner to your inner sanctuary. You don't have to know them yet, and remember we can't impose our will on another if you do know them. Sit across from each other. Look into each other's eyes. Feel a real connection. Bathe them and yourself and the relationship in the Violet Fire. Call the Law of Forgiveness to heal separation from Source. Decree the transmutation of cause, core, record, effect and memory of anything separating either of you from your highest potential and your gifts for the world and any discord. Speak through your Presence to their Presence and tell them of all the dreams you have. Now listen to theirs.

Step 9. Call to the Presence of your new partner, and charge their Threefold

Flame and the four earthly bodies with the Cosmic Light of the Presence and the sacred qualities you most treasure. Bathe in peace and harmony. Together create a symbol or mandala that represents your relationship. Trust what comes. When we work through the Presence it is not interference to call in the highest qualities for another. We can call these most important qualities for the entire world and assist in the shift of the ages. Our protection against accidental manipulation is to always decree it is for your own highest good and the highest good of all, which are never opposed. Assign an ascended master to your cause. Just call for the perfect one.

Step 10. What do you need to change in your life? Where have you dropped the ball? Stepped out of your power? Abandoned yourself?

Step 11. Seal it! Decree *The Will of the Presence is all encompassing and complete!* Bless it with peace and harmony to sustain it. Think of it as a garden. Seed it, water it, give it the sunlight of your Presence. Trust. Let go of expectation. Know it will manifest if you follow your guidance and take the steps it provides.

Step 12. Come back. Create a drawing of your mandala of love. Pin it where you can see it. Place the energetic template it represents in each one of your energy centers.

Step 13. Give gratitude to your I AM Presence. Acknowledge that at the I AM Presence you and your partner are one.

What to do

Be patient. Keep working on yourself. Send an arc of immortal love and the transmuting Violet Fire daily to your counterpart wherever they may be. Continue to work on finding your own autonomy, your sovereignty, your self-worth, your health and fitness, your joy and your purpose. Hold off on liaisons that have no meaning. Make a vision board of happy couples from the human and animal and angelic kingdom. There are some beautiful images available. Connect daily in meditation for direction. Bathe your vision with the most enthusiastic and purified emotion of gratitude and love that you can.

Each time you re-enact the steps of this template meet your partner in your inner sanctuary. Walk and talk together. Continue to work on yourself.

What not to do

Pick at yourself, be of two minds, talk about what you are doing to others. Lose hope. Inject anxiety. Fall prey to doubt. Tell friends and family what you are doing.

Remember the master's code. To Know. To Dare. To be Silent. To Do. *To know* is the acceptance of the full activity of the Presence and surrender to its wisdom. *To dare* is to have courage, which sometimes means healing wounds of low self-worth. *To be silent* is to not show the baby around before it is even born into the physical reality. You are creating powerful templates of energy and calling the forces of universal intelligence to fill the template with form. The force of others' doubt, opinions and emotions can destabilize it. *To do* is to take every action you are guided to.

By faithfully applying the techniques for manifestation in total trust, determination and commitment, you will change your inner world and then your outer world. You must transfer your attention to the inner world of the Presence and give no power to the outer world of appearances. All of these templates are dependent on increased life force and this is the province of the Presence.

On 'I AM Presence time,' you will create a new destiny and become a conscious creator, a creative alchemist, Homo illuminatus, an illumined one. Every time you use these methods you are creating an outline, a blueprint in the quantum field that draws more and more substance. The masters tell us that in the upper dimensions are many, many templates floating around abandoned just before they gathered enough force to manifest. Keep going and keep it light and joyful. If desperation and anxiety arise, go back to Phase One.

Practice active listening. You may be guided to *turn right, step into that shop* and end up conversing with someone who says you are the person they are looking for and what they are offering is the job of your dreams. Or you start talking about a travel book and end up having a cup of tea, and another, and three months later you are living together. If you override the still small voice within you will miss opportunities. Give yourself time, have patience. You need to take action and follow all the steps. You need to avoid activities that will interfere with your goal.

When you manifest from your I AM Presence you are in alignment, trust and surrender. You are in perfect equilibrium between action and non-action. Surrender the fruits of your labors to your I AM Presence with gratitude.

You can create your own templates for anything you wish to manifest.

Remember!

To Know.

To Dare.

To be Silent.

To Do.

Hold it all lightly and see it as a curriculum. Here's a review of some key qualities we are mastering. Understanding gratitude, forgiveness, love, power and wisdom better helps us put them into practice. As we master them in our lives our inner reality shifts, causing our outer reality to also move in seemingly miraculous ways.

Gratitude

Gratitude is foundational to any practice. I have come to realize that it is one of the great mystic keys for opening the treasure chest of the Presence. Gratitude brings grace and restores the natural order of things. It also expands that for which we are grateful and is so potent it could be an entire practice on its own. The clearer you become the more natural it is to be in a continual state of gratitude.

Try to be grateful for everything. Spend some time every day giving thanks for your life, home and circumstances, even if your current circumstances are troublesome. See troublesome circumstances as learning experiences. Be grateful for your parents no matter what your childhood was like. They got you here. Be grateful for all generations proceeding you and following you. Be grateful for where you live, friends, food, and so on even if life isn't exactly where you want it to be yet. When we are in continual gratitude we are open and receptive to receiving.

Every day I give thanks to my Presence, my Higher Mind, my four earthly vehicles, the universe, our Earth and all its realms and beings, all directions and elements, those who guide us and for every electron, particle and wave holding the forms in my life. I even thank my appliances. I have found gratitude as a practice to be life-changing. I am better able to see what I do have and the kindness, ease and grace in my life multiplies yearly.

Forgiveness

Scientific research has confirmed that forgiving yourself and others will create a greater spike in consciousness than almost any other practice.[188] All wisdom traditions speak of forgiveness as a key tool for maintaining inner peace, which is necessary for succeeding with manifestation. It sets you free. Forgiveness heals separation and restores unity.

What exactly is forgiveness? And what are its mechanics?

The first and hardest is when we are caught in the perceptual role of being a victim. Here, we still see reality as happening *to us.* We're still playing a game of projection, blame and shame. No doubt we are angry. We are also drawing angry situations to us in one guise or another. At this level of perception forgiveness seems like a bitter cup. In fact, it is often imposed and insincere, anger still roiling below.

The second level of forgiveness occurs during the emotional clearing steps of Creative Alchemy if application is diligent. When you have released the original buried emotion, which has created a self-sustaining pattern, the pressure goes and compassion flows. You can begin to more easily forgive yourself and others. It feels really good. You realize your lack of forgiveness has been harming *you* most of all.

The third stage is the realization that forgiveness heals the illusion of separation. No one can do anything to you. Everything is a mirror, a consequence of your energy field and its attractions and aversions. You create your reality. Don't expect to leap to this stage quickly and many people recoil at even the thought of it as it brings to bear many sociological questions. However, it will be a perception that is finally reached. As the saints from India who would call our world Māyā, 'a constantly shifting dream' or Lila 'divine play' and Shakespeare said, 'All the world's a stage. And men and women merely players,' we begin to see life as an intricate dance. In our mastery, we cannot blame so how can there be anything to forgive? We are the creators.

The fourth stage is when we accept the entire reality that we inhabit is an extension of our consciousness and we call upon the Law of Forgiveness on behalf of the whole. We can bathe the world in Violet Fire and call for cause, core, record, effect and memory to be transmuted into Immortal Love or Victorious Peace or Cosmic Victory.

Each stage of forgiveness arises naturally from the previous one. As we release the buried trauma that causes us to be triggered and to recreate patterns of projection and blame, it is much easier to forgive. If you can achieve levels two and three you are doing well.

If you feel you are stuck or resistant, keep applying the exercises in Phase One of Creative Alchemy. These will enable you to work through whatever has taken hold of you emotionally. Be diligent and committed until it is released. It is so worth the effort. When your mental 'projector' is cleared of the patterns of blame it is freed to begin projecting pure vision.

Love

The English language has few words to describe the various states of being and feelings we generalize under the term *love*. By comparison, the ancient Greeks had eight words or more for love. We use the same word to describe our delight in a new sweater as we do to describe how we feel about those closest to us.

There are many types of love, beginning with *the love we have for our mothers* as small babies, which grows into the love we have for our other family members and friends. This fundamental love is crucial to the human race. When we get this right, other iterations of love emerge from it.

Then there is *self-love*. Not narcissism, but genuine adoration and love of self. Really, we are born with this. But most of us have to go hunting for it later. Our cultures for the most part don't understand it. We are told not to boast, not to get above ourselves, and so on. Why shouldn't we be proud of what we have made honest efforts to achieve? And why shouldn't we share that? Not to lord it over others, but to celebrate. You can only become the best you can be and no one else can be you! To feel the love inherent within the consciousness from which we emerge, we must first love ourselves. Self-love is an antidote for the sense of abandonment many of us feel. We can essentially become our own parents. Go on, adore yourself! This love forms the basis for self-care which is essential to our psychological and physical health. We are love. Love is the cohering force of the Universe.

The *romantic love* of two people for each another can be expansive and supportive and add meaning to our lives. But when our emotions are unhealed, romantic love can lead to jealousy, blame, obsession and possessiveness. When you love yourself, and want to share your life with another person and when challenges occur, are willing to hold the mirror up as a reflection of what is unhealed within, a healthy relationship is created.

From romantic love emerges *Eros,* love expressed sexually. When two people are whole and healed, their sexuality is a sacred act of coming together and engaging mind, body, heart and soul. Two loving beings merging in this way is a transcendent act. When it dwindles to a mindless entertainment, an addiction, a distraction or a way to fill a pit of loneliness, it subtracts from us. Sex can be a sacred, intentional, illuminating, transcendent activity. Why not experience the full sublime potential of love expressed as Eros?

Compassion is another form of love. When our hearts are clear and we've done the work to release our suppressed emotions we naturally ignite embers of compassion in our hearts. Compassion is tenderness, charity, benevolence and caring for others. When we intend to love in this way we light the spark within us and fan its flame in others.

Empathy is an element of compassion. It is an act of the imagination by which we can imagine what another person is feeling. It is stronger than sympathy. Pity is a judgment. Empathy and compassion are ways of feeling for another without making a judgment on their circumstances or how they got there. Things are rarely as black and white as they first seem.

When we begin to know ourselves in our true essence we naturally flow closer to *unconditional love*. It is important to understand the difference between emotional love and unconditional love. Emotional love is only stable to the degree our emotions are healed. If our emotions aren't healed, what we think is love will actually be needs and desires motivated by a feeling of incompleteness and therefore inherently unstable. There's a lot of confusion about unconditional love and I was certainly confused myself for a long while, thinking it meant opening my door to the world and caring for everyone who came my way. My lack of discernment led to some sticky situations. When we are in alignment we know when to help and when helping is actually meddling and enabling, harmful to ourselves and others. Unconditional love involves non-judgment and compassion. It is synonymous with Radical Acceptance. You will experience this kind of love when you stop comparing and measuring others – for judgment is always a projection. Using the guidance of our Presence we know what steps we are meant to take to help others.

Unconditional love inspires us to serve others and the greater whole. This desire to serve the greater good is inherent in all humans no matter what their beliefs are and it becomes suppressed when we experience trauma. It is our natural healed and whole state and the source of great joy. In fact, the desire to serve the greater good is present in animals and thanks to new research we now know plants as also serve their community.

If you're not in a romantic relationship, why not redirect your sexual energy and love into *creative expression*! It's your life-force energy. Use it well and enjoy a long ebullient life. True authentic creative expression is an act of love. When we create from our hearts it is one of the most healing acts we can engage in. True authentic creativity aligns you with the ultimate Source. It is the same force that creates the beauty of nature and fuels life. Engaging in deliberate acts of creativity can lead to wonderful experiences of inventiveness, playfulness and inspiration. I call it *devotion in motion*. It's evolutionary and revolutionary.

The next stage of love is *limitless creativity*. Inspiration is pouring from us. We start to reflect the creative evolutionary Source we come from and we see, feel and know this level of creativity is based in love. We are magnetizing invisible potential into visible, tangible, manifested form in space–time reality. There is sacredness about it because it has higher purpose. We begin to receive inspirations that will bring benefits to ourselves and others.

Then there is the *cosmic love* inherent in the universal field. This love can really only be experienced. There are no words to do it justice. It is the recognition of a drive towards coherence and of the harmony behind all objects and wave states. It has magnetic properties and is the ultimate power that holds form in place. In the East, universal love, *Satchitananda,* is the experience of the unchanging reality behind all forms. It is pure existence, consciousness, bliss. It is a fundamental truth that the magnetic cohesive force behind all things can be called love.

Love is an awesome force of extraordinary power. It is *the* force.

True Power

True power is a build-up of energy that has been refined by living with integrity, distilling knowledge into wisdom and being in harmony. To build true power you must use your energy impeccably. This means not squandering it, abusing it or using it to control others. Be aware of self-pity, recrimination, holding onto past slights, which consume your power. The moment they appear, go back to Phase One.

Power can be built by certain practices such as Qigong. The Fire Breath used in Yoga is also excellent as are certain forms of Kriya Yoga. Self-Realization Fellowship (SRF) teaches the Kriya Yoga technique Paramahansa Yogananda brought to the West in the middle of the 20th century, passed down from a lineage of avatars, self-realized beings. Among other things, it centers on a special breath that moves along the spine to the third eye, dissolving obstructions. Yogananda called it the 'airplane method to God' (pure aware-ness). The great Gurus of the lineage become our inner-plane teachers when we learn Kriya Yoga through SRF.

True power is also the ability to sense, read and see energy and the ability to direct it. This can also be learned. You begin by intending, focus and attention. Aside from the power we build we inherit power and are born with varying amounts depending on lineage, conception and our own accumulation in the Causal Body.

Power can also be stolen. In our hypnotized world people are attempting to steal power from one another all the time. Use your amazing physical body for discernment. If a situation doesn't feel right withdraw until you understand why. Needy people suck power. Never take energy, ideas or any other thing from another. Remember your sustaining power is the Presence within.

Remember to reinforce your force field every day and every night. Calling down the tube of white light from your I AM Presence every day is good practice, as is surrounding yourself with the Violet Flame to dissolve any discord coming towards you or leaving you unconsciously.

When you go to sleep ask to be taken to the temples of learning appropriate to your goals.

When you are healed and clear, ready to learn the art of destiny and practice of the science of miracles, the more power you have built up and learned how to utilize, the easier it will be. You will discover that true power fills your cup to overflowing and naturally inspires service to life.

> *I slept and dreamt that life was joy. I awoke and saw that life was service.*
> *I acted and behold, service was joy.* [189]
> – Rabindranath Tagore

Measuring Power

I was at Grimstone Manor in Devon, England, studying with the Nagual Dreaming shaman Merilyn Tunneshende. We were learning how to increase power through dreaming, special conscious lucid dreaming which entails crossing into other planes of existence, other worlds. 'Dreaming awake.'

She was talking to us about power. She brought out a very special Brazilian Rain Stick. A Rain Stick is a natural instrument made of a hollowed-out piece of wood or large hollow reed filled with seeds. When you turn it upside down it makes a beautiful sound, the sound of light rain on broad leaves, a river of gentle taps.

Merilyn passed around the Rain Stick and told us to turn it. The seeds would fall, making the sound only as long as the measurement of each individual's power would allow. Amazingly, the Rain Stick responded differently to each according to their level of power. One man turned it once and then the second time there was no sound. Merilyn said he must spend this life building his energy. Others had varying results. One person turned it eight times and it was still flowing when they handed it on, a little embarrassed by their degree of power. It was a fascinating experience and very salutary. Energy is how we manifest.

What happens when we squander it? Can you sense your inner power? Are you aware of it? Could you give some time to building it? Could you become very careful with how you spend it?

Wisdom

Wisdom is a distillation of inner knowing, experience and knowledge. We can have several PhDs and have read every book in the world and speak ten languages and not be wise. We can't actually force becoming wise but we can intend it. Silence helps to develop wisdom. Thinking before we speak and generally speaking less. Meditating helps us connect with deeper wisdom contained within our hearts and Presence. We gain insight and flashes of revelation that helps us see patterns and a bigger picture than we did before.

Study and application followed by intense relaxation lead to 'Eureka' moments where a whole new level of perceiving is revealed.

Kindness is the greatest wisdom.

Summing Up – The Alchemy of Creativity

Creative Alchemy is conscious creation of our destiny, our life. We are always creating, but most of us create unconsciously, coloring our manifestation with the unhealed *emotional wounds* of our psyche and stuck patterns of *identity.* When we have healed our memories and emotions, we become clear vessels endowed with the ability of *fresh perception,* uncolored by patterns of the past. Our imposed identity falls away to reveal our true essence where the keys and codes of our highest purpose await us.

> *And above all, remember that the meaning of life is to*
> *build a life as if it were a work of art.*
> – Abraham Joshua Heschel

- The power to *imagine* is a function of the *mental body*, your quantum manifestor. Remember, patterns created by thoughts that are indistinct and imperfect, or much worse, emerging from latent fear or hate fueled by unhealed emotion, create chaos. Once you have cleared your emotions your mental body calms. New neural pathways are laid down and positive chemicals and hormones are released. You can begin to train your mental body so it is the receiver of power and not the doer. It can unburden itself and relax and fulfil its purpose to receive the guidance, take the necessary steps required, analyze, measure, focus and create the detailed architecture of your dream using its ability to imagine.

- The power to *feel,* the function of the *emotional body*, your quantum generator, fuels the pattern or template created by the imagination and creates *Living Vision,* a function of the Higher Mind, the bridge between our mortal self and our immortal self, the Presence. When your emotions are healed and no longer suppressed, enormous power is unleashed.

- Entraining your force field to resonate with the pattern and tone of your Living Vision is a function of the *energy body,* your quantum conductor. The records of discord imprinted in our subtle energy body can be cleansed and healed by ardent and committed use of the *Law of Forgiveness* and the alchemist's ray, the *Violet Fire.* Just as our physical body is cleansed by soap and water, our energy body can be cleansed by the application of heightened frequencies, which absorb lower ones.

- The action and steps you must take following the guidance of your Higher Mind is a function of your *physical body,* your quantum navigator. Care, love and strengthening of the physical body is essential for success.

- The final spark of life and animating electricity is bestowed by the inner superhero, the *I AM Presence* magnified by the science of decree. The activity of the *Threefold Flame* in your heart releases the *power* necessary to consciously create and manifest in the physical, the cohesive and magnetic *love* to sustain it and the *wisdom* necessary to create from inspiration. Your intent that it is for your own highest good and the good of all insures there is no misuse of life force. There is only one power that can act, that of the Presence. Your free will allows you to create discord or harmony with this pure life force.

- The power and substance of *conviction* holds unshakeably the belief that your Living Vision will emerge into physical reality. The quality of *harmony* gained by the healing of emotion and alignment is the key that opens the floodgates of the substance of abundance you will need from your *Causal Body.* The quality of gratitude creates the imprint of already having received your dream. The quality of *peace* sustains it. This process gradually becomes seamless.

- Withdraw your attention from the outer reality and stay focused on your dream. Reverse-engineer when necessary to see what needs attention and clearing. You will find yourself in greater and greater surrender to the inherent pattern for your life, the one that is perfect for you, imbued with beauty and meaning and purpose and wholeness.

- Now to plant the seeds for your beautiful garden to blossom under the radiant sun of your Presence. Your future is golden. Your future is boundless.

Part Three examines how we can apply some of the principles of Creative Alchemy to business, education, families, communities, mental health, art and co-creating a new world.

PART THREE

THE BROADER APPLICATION

O! for a muse of fire, that would ascend the brightest heaven of invention.
– William Shakespeare

How do we apply the practice of Creative Alchemy to create a new worldview, a new paradigm that will begin to emerge as a new world and a new Earth and make a paradise of our Earth? What would our world look like if everyone had access to illumined inspiration and heightened solution and innovation abilities? How quickly could we solve our problems if we were in alignment with the very Source of life, united in inventing holistic systems for our planet? How efficient could we be if we worked together on a planetary scale shifting planetary patterns at the core? What would it look like everyone understood their supply was within? They did not need to steal or hoard?

Part Three explores innovations and paradigm shifts already happening in the world and proposes ways we can practically apply Creative Alchemy in business, education, prisons, parenting, grief counselling, mental health and art. It also examines how outdated the anthropocentric worldview of the human as supreme planetary intelligence has become. When we truly understand that the same consciousness permeates and animates all life, our well-being and the health of our planet will reflect this.

The importance of creativity in the next step of our evolution is incalculable for all life on a personal and planetary level, as is the importance of harmonious collaboration. The more we know about global coherence fields the more we know the personal is the planetary and the planetary is the personal.

Those who understand the role of emotions in our creativity and the role of creativity in our well-being and who see the ability to innovate and access transcendent imagination as crucial skills for our continued evolution and development are designing a new world where we will play our work and create new systems in all areas. These include ecologically and socially balanced financial prosperity, advanced solutions for planetary care and clean-up, solutions for the healing of trauma, an emphasis on emotional literacy in education and business, the long and slow reinstatement of the disenfranchised without diminishing cultural heritage, a revaluing of the education system to embrace the full scale of human potential and planning the cities of the future. The evolution from Homo sapiens to Homo illuminatus, the visionary human who understands the principles of manifestation and collaborative co-creation, is fully underway.

7

CO-CREATING A NEW WORLD

True creativity and innovation
come from standing at the interface of art and science.[190]
– Walter Isaacson

Creative Alchemy for Business

The companies and individuals who are able to adapt with speed *and innovate* are the ones who will rise in the coming years as important local concerns and also as the new world players. Those who have not accessed the full weight of their creative potential will be dragging their heels.

AI will soon surpass us in abilities to compute and administer – think of IBM's Watson, which can make associations between 40 million medical documents in 15 seconds. We can embrace this technology and enhance it by radically upgrading our creativity, positive solution finding and imaginative innovation. Intelligence will not be measured by knowledge which any computer will be able to provide but by imagination.

When our own visionary potential is unlocked, we can unlock the creative potential of our teams and colleagues. The next frontier for humanity is heightened *authentic* creativity, visionary thinking and *collaborative* co-creativity.

Remember the studies by neuroscientist Cathy Stinear that I mentioned earlier? High-level creatives have brainwave activity flooding back and forth across the brain, collating and collecting, much like gamma waves, our doorway to other dimensions. She and her colleagues also discovered that less creative people have a brain function that moves up and down, not back and forth. However, when two individuals were put together their combined functions were equal to or surpassed the higher-level creatives, highlighting the importance of teamwork and brainstorming. We now know the incredible plasticity of the brain. We can learn to be more creative and flexible thinkers. Our neurons and their connections continually change, based on use. As Stinear states, our brains were built for creativity.

The point is that teamwork in business is crucial. Brainstorming with an entire team magnifies potential and creates inclusion that in turn inspires loyalty and motivation. You never know who is going to have the light-bulb moment.

On this note, it's crucial for a business not to stigmatize failure. This puts the cap on innovation more surely than anything else. Richard Branson states he doesn't even want to meet with a businessperson who hasn't failed several times. He's not interested in safe; he's interested in risk-takers who push the boundaries.[191] You don't have to be mind-blowing, just allow a margin of failure in direct proportion to the degree of innovation you'd like to see. I think failure should become an obsolete concept anyway. Let's just call it another learning experience! Some schools now present the concept of FAIL, which stands for First Attempt in Learning, a very healthy concept.

The boss is another outdated concept, borrowed long ago from the military. Now you will know who the leader is by noting the one who is inspiring the team and finding ways to accelerate purpose, inspire motivation, shift perspective to holistic goals and encourage teamwork. The new paradigm 'boss' will be the one concerned with innovating new ways to stay relevant. They will be creating growth-oriented culture that is ecologically responsible, promoting emotional literacy and eliminating resistance to change. The priorities are shifting towards finding ways to access personal and group vision, analysing the past with a view to learning and creating the future. New paradigm businesses are measuring all goals against the higher purposes of planetary goals. Ecologically minded businesses are getting a lot of attention and rising above competitors. Ecological investment portfolios are gaining appeal and Green Investment Management is a growing sector.

Personal evolution is crucial if the companies and businesses we own or work for are to evolve. Paradigm shifts for business include new models of sustainability and cooperation over competition. Customers and investors are increasingly voting for ethical business with their loyalty and money.

> *One picture is worth ten thousand words.*[192]
> – Frederick R Barnard

Business is learning that, in order to grow, the people involved with running it need to grow. Understanding the role of emotions in communication and creativity play a big part. When we approach work problems as play with emotional fluidity, we will inevitably be more creative. Thought leaders and creative facilitators such as Elysa Fenenbock, Google's first designer-in-residence, are changing the way we approach business by drawing on techniques that encourage creativity and the understanding of emotion.

Elysa, whose Creative Nomad Project leads global organizations to foster creativity in their endeavors, states that CEOs are realizing that creativity is the single most important quality for success in business.[193] Innovation, imagination, emotional empathy and connection are the new buzzwords in the world of business. At a Creative Leadership conference that I attended, Elysa Fenenbock had participants form teams and build visual sculptures based on an agreed emotion. The other teams had to try to work out what emotion was being depicted. She believes creating emotional empathy is crucial to a team's collaborative abilities and also its ability to innovate and create a future vision for the company.[194]

Creative people perform in heightened uncertainty and thrive when innovation is called for. They like to surprise themselves, take risks. Applying the habits of artists to business helps create solution-oriented environments that find answers for challenges that haven't yet presented themselves.

Aithan Shapira is the founder of Making to Think, an organization that fosters breakthroughs in leadership and strategy using the arts. He designs experiences to shift perspectives in how we see and hear in order to expand and promote team collaboration. If our emotions are free flowing, we will be less rigid and able to shift our belief structures and perceptions as needed to address new situations. Shapira also uses playful exercises to illustrate the role of creativity in business. A live chamber orchestra playing the same piece of music in three different ways at his workshops challenges our skills of listening and other perceptions.

Gus Balbontin, former executive director and CTO of Lonely Planet and technology, creativity and team performance advisor to GoogleX, Nokia and Amazon, reminds us how critical this adaptability is and how little time we spend addressing it. He currently runs a design and innovation agency and is Entrepreneur-in-Residence at Victoria University of Wellington, New Zealand, analysing what companies did right and wrong in these last few decades of unparalleled change, staying close to large corporate and small business challenges. He points out that predicting trends and influencing technology is less important than being adaptable to what is coming, stating that most useful lessons for life and business happen in the school playground. He advises balancing momentum with taking stock for re-invention, which he believes is crucial for survival in a lightning-paced changing world.[195]

Business leaders are also understanding that creativity is the process of having original ideas that add value. The techniques distil high-concept ideas into easily accessible modules for an on-the-go audience. Someone who has unleashed the power of their healed emotions is that much more creative, imaginative, solution-oriented, highly functioning and also more in alignment with the highest good. Observation and reflection are a necessary part of the formula as well as inner accounting. Are there emotional connections in the team needing to be cleared? Are there doubts and fears interfering with the strength of mutual intention? Is competition interfering with collaboration? Are the right steps and actions being taken? If not, you will build a nice castle in the sky, become derailed or fall short of your vision. The process of reflection and adjustment can be used to analyze manifestation. Does your intention need an adjustment? Working backwards from that, check in on the current beliefs, perceptions, thought patterns and emotions of the team. What actions have been taken? Has one been left out, or has the approach become overly detailed or sloppy? Is someone having trouble delegating? Applying the steps of Creative Alchemy has been successful in opening creative faculties, creating trust and bonds in teams by sharing vulnerability and support, opening imagination and understanding the mechanics of manifestation.

A group of people holding a vision and taking the correct steps together is a powerhouse. When it is also for the highest good, the power increases exponentially. Having a strong mission paper is crucial to creating powerful team focus.

The rise of B Corps, short for Benefit Corporation, is also changing the landscape and addressing the need for change. B Corps-compliant, for-profit businesses become committed and legally obliged to be a force for good. The four main tenets of being a B Corps-certified business, from their Declaration of Interdependence, are: that we must be the change we seek in the world; that all business ought to be conducted as if people and place mattered; that, through their products, practices, and profits, businesses should aspire to do no harm and benefit all; and that to do so requires that we act with the understanding that we are each dependent upon another and thus responsible for one another and future generations. There are currently over 2500 certified B Corporations in more than 50 countries.[196]

More individuals are interested in businesses that generate jobs and improve society. Sixty-four percent of millennials stated making the world a better place was their main priority in a study from the Intelligence Group, and 88 percent wanted 'work–life integration.'[197] The idea that maximizing profits is the main goal is becoming a dinosaur. More of us are voting with our wallets, shopping at, seeking employment with and investing in ethical, socially and ecologically sustainable business.

Ecologist, physicist and systems analyst Fritjof Capra, who wrote the groundbreaking *Tao of Physics* in the 1960s, proposed 'caring capitalism' in that book, a holistic way of increasing overall economic well-being. For example, if large-scale pollution is allowed for business because that boosts economic growth, what does that look like if measured against the costs of ecological clean-ups and associated health-care costs? Capra has further developed these ideas, producing a conceptual framework for economics based on the systemic principles of life, concluding that the purpose of the economy is to serve the life purpose of social and ecological systems.[198] He states there is a need for a 'new transdisciplinary economics, which unites opposites and creates a basis for peace in ourselves, between people and between people and nature. Today's economy generates conflict in all areas.'[199]

Like the need for war, rampant uncaring economic growth is a reflection of internal conflict and unhealed emotions.

Creative Alchemy for Schools

Creative Alchemy in every school would change the world in one generation. I have taught Creative Alchemy to schools and children in the United States, England, New Zealand and Australia and children respond immediately and deeply. It's extraordinary how big the emotions of little people are. They often begin not even knowing the difference between a thought and an emotion. It is unlikely their parents or teachers know they are feeling huge emotions, often of despair, rage, hopelessness, anger but also joy, peace and calm. It is wonderful seeing children intuitively transmuting their emotions without forcing it. Another emotion often comes through and then another and gradually the alchemy occurs. Children are delighted with learning a process that gives them tools to understand and handle large emotions.

Once explained, children easily grasp that an emotion is something they can actually feel in their bodies and they can learn to sense the feeling and understand it. No emotion is called good or bad, they are all friends if they move freely. Stuck emotion becomes *E-Motion*, energy in motion through Creative Alchemy.

Teachers are often tense and want their students to paint pretty pictures. Usually the child I am told can be 'put outside' if they are disruptive is the child with the most pent-up emotion. I always have the teachers do it as well, so they get a deeper understanding of the process and understand it isn't at all about pretty pictures. They are ultimately very pleased with the introduction of Creative Alchemy into their classrooms. Children become more focused which leads naturally to better results and there is inevitably less disruption and better communication.

Children find these exercises hugely relieving. I have had very moving experiences such as a little boy with OCD coming up to me after a class to say he felt relief for the first time, or a pre-teen telling me she is relieved of tumultuous inner anger and anxiety for the first time. Teachers tell me children diagnosed with ADHD also become calmer. Children learn that the safe communication of big emotions leads to the release of repetitive negative stories that hold them in melancholy or stressful states. Children who are physically challenged often release huge rage. Their smiles after workshops are testimonials to the relief they feel.

Separate Creative Alchemy workshops for teaching staff are also effective. Teachers are under a great deal of stress and powerfully respond to the exercises. Sharing vulnerability strengthens team spirit. Teachers are undervalued and overworked yet they provide one of the most important roles in our society. All of the teachers I have worked with accessed and released emotion and communicated with their bodies. Many of them received profound messages from their own bodies for chronic conditions. Many teachers seem to have chronic lower backaches, a sign of feeling unsupported. Often things are stirred within adults in a group situation that may need more time – the first viewing of a buried trauma, a flash of deep insight accompanying the emergence of a long-suppressed emotion.

Our education systems need to be overhauled to assist the shift from Homo sapiens to Homo illuminatus. Inaccurate histories are taught solely from the points of view of the conquerors. Our science education is often outdated. The entire mode of rote learning still used regularly in some parts of the world is antithetical to training minds to be innovative, solution-oriented, creative and inspired. The factory model of most schools emerged in the late 18th century and hasn't changed much, although there are many glowing exceptions. As AI quickly races forward with extraordinary computational abilities, we need to increase our faculties of creativity and imagination. Teachers need to be valued so they can be impassioned to light the spark in the next generation. Here is a list of thoughts of how we can change the world in one generation by changing our models of teaching and how we teach. It is so short-sighted to undervalue the education of our children. They are our future.

Changing the World in One Generation

In a Tibetan monastery, an 11-year-old boy discusses physics and the nature of reality with a Buddhist monk and a MIT physics graduate. In the Chicago ghetto, young children from severely disadvantaged circumstances eagerly discuss Shakespeare and Chaucer. What is the difference between these children and the many children who are leaving schools with few skills, little confidence or sense of purpose? In the two above examples it was extraordinary teachers who made the difference. But for teachers to maintain enthusiasm and dedication they need to be honored. So, that's first on the list.

1. Honor teachers. Have awards for excellence that are nationally recognized. Prioritize good pay.

2. Teach Creative Alchemy in all schools so children may learn to be emotionally fluent and literate, understand their true inner power, their ability to manifest what they can imagine and connect to their inner Presence, inheriting their true birthright. Through the techniques of Creative Alchemy they will gain the skills to co-create a new world of inspirational collaboration, sustainability, innovative problem-solving and cooperation without war, which is an out-picturing of inner conflict. They will gain the understanding that they can create great health, prosperity and joy through their inherent abilities.

3. Elevate the education of all children to the top social and political priority.

4. Encourage luminaries from all walks of life to give time to their local schools, offering inspirational talks and mentorship where possible.

5. Ensure all children are exposed to the arts and have a rich creative curriculum.

6. Teach comprehensive holistic overviews. What are the connections between lack of funds for education and youth crime? Between pollution and health costs? Children are brilliant. Let's not talk down to them. Encourage invention, innovative problem-solving and creative thinking. If children are taught the principles of inventing they will innovate solutions we haven't yet dreamed of.

7. Apply an interconnectedness between disciplines such as studying the relationships between mathematics and music, law and logic, geometry and the ancient science of sacred geometry, between human and natural proportion, between theology, metaphysics and quantum theory, between religion and philosophy, between energy, sound, color and light etc. Renaissance thinking encourages a holistic worldview and an under-standing of the interconnectedness of all things.

8. Build visual representations of culture. Get off the page.

9. Combine art and history and geography to make all three come alive. Create a theatrical production of the era or nation being studied, listen to its music, study its art, make art similar to the art of the era or place, cook its food. Bringing cultures and places alive in this way engages the whole being and does much more than teach, it literally awakens a connection to a foreign culture in a visceral, meaningful way. While learning the exciting differences, children also learn the commonalities. Teaching the similarities

within world religions and cultures and celebrating the differences would go a long way towards removing racism and cultural prejudices from the planetary group mind.

10. Teach meditation techniques. Begin with a simple technique such as the HeartMath meditation for children, then move to the Creative Alchemy Meditation and the I AM Presence Activation. The Presence of humanity is united and beyond all religious separation. Statistics from the David Lynch Foundation for Transcendental Meditation in schools show a 21 percent increase in high school graduation rate, a ten percent improvement in test scores, increased attendance and decreased suspensions for high school students, a 40 percent reduction in psychological distress including stress, anxiety and depression, reduction in teacher burnout, reduced ADHD and symptoms of other learning disorders and increased intelligence and creativity.

11. Teach yoga that is meditation for the body, or stretch or dance. Elkins Elementary School in North Texas increased their recess to four fifteen-minute periods a day and found the children no longer fidgeted during classes and were able to maintain total focus. They were inspired by the LINK Project, a wonderful initiative which stands for Let's Inspire Innovation 'N Kids. They are working on other initiatives to build character and shift procedures to include the whole child.

12. Teach sports that focus on team spirit and develop physical and psychological skills. It is said the gods gave us sport to distract us from war. Children who are sporty and strong are more confident. Learning to win or lose with grace is a valuable skill for life.

13. Encourage physical expression such as singing and dancing for sheer joy and release of physical tension. This creates enthusiasm and well-being. Enthusiasm is the antidote for fear. Teach that life is a grand adventure, not a series of boxes to be ticked. No child should be excluded.

14. Teach informed debate.

15. Children should not be patronized. However, firm structure provides a sense of security and is the basis for personal self-discipline. Teach courtesy and respect.

16. Explain the reasons for rules and regulations. This models respect, articulation and clarity. Children are less likely to rebel if they understand there is a good reason for a rule. People who can articulate are much less likely to express themselves violently.

17. Teach classical literature as well as new work. Create bridges rather than burn them.

18. Help children find their personal gifts.

19. Teach gratitude, forgiveness and compassion.

20. Teach children they have the ability to draw the energy from their electronic Presence and heal themselves and others. Children respond very strongly and quickly, as they don't yet hold negative belief systems about their capabilities.

21. Instil social values by trips to local homes for the elderly or visiting children in cancer wards.

22. Instil the philosophy of Earth Jurisprudence, the philosophy that we are part of a whole and the welfare of each member of that whole is dependent upon the welfare of Earth as a whole.

23. Have organic school gardens so children learn to grow their own food, make healthy juices and understand the basic principles of a healthy diet. Rooftop gardens or window gardens can be created if there is no available land. Seeing how things grow is inspiring for children.

24. Teach the basic practicalities of life: how to make one's own clothes would encourage personal expression. How to cook healthy food, change a tyre, open a bank account and so on.

A note on teenage suicide, which according to World Health Organization statistics is growing worldwide. I had the honor of hearing an esteemed Aboriginal elder speak on the subject and offer the best ideas based on her own astounding results. She had gathered all the children and teens in her community – there is a high suicide rate amongst the Australian Aboriginal people – and excluded the parents. She started meeting weekly with them for a few hours and guided the group in spiritually connected activities in a playful way. They created totem animals, dream catchers and so on. While they created, they spoke casually about their lives.

Remember, acts of creativity suppress the parts of the brain that suppress emotion, thereby aiding in the release of emotion. Authentic creativity is alchemical and has the ability to transmute. The suicide rate in her community went down by 100 percent. She believes these at-risk young people are highly sensitive souls who feel lonely and isolated in a confusing world. These simple exercises connected them to their physical and emotional bodies and to their Higher Mind where they could begin to receive the inner guidance they needed. The young community of like-minded people healed their loneliness and sense of isolation. Community is essential to our well-being.

If there is a teen or anyone in your life (or you!) who is feeling withdrawn and struggling, it is important to seek help.

Creative Alchemy for Parents

I had a lovely moment when a local friend told me her communications with her 12-year-old daughter had improved exponentially now that she had started to draw her emotions to express her feelings. I kept it to myself that her daughter had been in a group I had recently taught but felt inwardly happy that she had applied practically what she had learned and positively increased her communication skills.

There could be nothing better than teaching your children the steps in Phase One of Creative Alchemy as a way to heal and express their emotions, understand their power and help communicate strong feelings. Children are completely able to learn all the steps in Phase Two and Three as well. It is wise to teach them these techniques so that they are always in touch with their inner guidance, inspiration and discernment, and understand they have access to great personal power to be expressed through the lens of love and wisdom. They have no problems with big concepts.

Children who have experienced major trauma can be supported by these techniques but may need the support of a professional. If parents introduce the basic Creative Alchemy exercises at a young age they will find an ease of emotional expression and communication while also encouraging children to develop their creativity, innovation and solution-finding skills. My own beloved eldest granddaughter has been on this journey with me her whole life and she is exquisitely creative and emotionally self-aware. I would love for all children to have these benefits.

A wonderful exercise to do with small children and children of any age is ask them to envision their perfect world. It is the most charming exercise imaginable as children create a vision of their perfect beautiful world filled with lovely ideas. This simple exercise is the beginning of manifestation techniques. Now that we understand we are living in a great big radio of sound waves and the song of those waves can be shifted by concentrated intent, what would happen if thousands of us gathered together with a new healed template for our world? Millions? The science of Morphic Resonance teaches that only ten percent of a population is needed to shift the worldview of a group. I'm holding that vision. Please join me. Right now, that number we need to awaken is approximately 800 million, creating a seismic shift that would jolt the remaining people out of hypnotic slumber.

Creative Alchemy for Pregnancy and Childbirth

Creative Alchemy is excellent preparation for pregnancy and childbirth. Pregnancy, childbirth and the early years of our children often trigger the release of painful buried emotion. As these emotions rise up, they need safe expression otherwise they can become triggers, causing us to overreact. If they are released we are much better parents and guardians. The techniques of Creative Alchemy should be done by both parents prior to conception but begun anytime to reap the advantages for your pregnancy, family harmony and the benefits for your child.

If you or someone you know is pregnant or contemplating pregnancy, Creative Alchemy is an important tool. It will clear buried emotion and expand consciousness, which increases frequency and vibration. This expanded frequency will allow more of the child's essence to enter its Quantum Navigator, its body, and give it a huge head start in life.

Here's a little story of the miracle of childbirth and the consciousness of the unborn child. But as we are coming to know, all miracles are within natural law.

Communication Before Birth

My husband and I had discussed baby names all summer. I was very saddened therefore when he came to the strong decision that he didn't want a second child. He adored his stepdaughter with all his heart and one child was enough. On the other hand, I had felt a child's soul near me all summer long and wanted my daughter to have a sibling. I dreamed of the child and received messages from the child.

My husband remained firm and I reluctantly let go of my dream for a bigger family.

One day in my daily communication with my Presence I asked for some confirmation that I wasn't just talking to myself. I wanted information I couldn't have dreamed up. I received a message from a beloved guide saying the child that would have been born from me would be born from a woman I had met at a gathering in America. I remembered her well, a lovely, gracious woman with whom I made a strong connection. She had given my young daughter, who had accompanied me to the gathering, an exquisitely beautiful necklace of a dolphin with a child sitting astride it.

A few weeks after receiving this very surprising message, the woman called me one day out of the blue. She said she was in the Yucatan and she was pregnant. Her parents no longer were alive. She asked if I would be the guardian of her child should anything ever happen to her. Silently my jaw dropped. I said of course I would. I decided not to tell her the full story of the message I had received. It was too strange!

A month later she called again. She had bought a book called How to Communicate with Your Unborn Child. *This was going very well and she was receiving lovely and surprising messages from her child. To her surprise she had received a message that I was also her mother and my daughter was her sister. At that point I was compelled to share the guidance I had received.*

My daughter and I went to the Yucatan and brought her back to London. Once we were back in London we saw her only occasionally. I was not having an easy time of it at home and had withdrawn. This was around the time of my radical awakening, and a lot of unhealed emotion was surfacing and the dynamic with my husband had also suffered due to his decision. Being charismatic and outgoing, she quickly created a life for herself. She had a large circle of new women friends and a select group would be at the birth. I was excited to meet this baby who may once have planned to come through me.

One night I received a call. Mysteriously, no one in the birthing group was available. I would be the only one with my friend when she gave birth. At the hospital my friend began to tone like a glorious sea goddess. No one at the hospital seemed to mind. Everyone was rushing around and overly busy. I got her into the birthing tub at her request. The midwife poked her head in and said to me, 'Keep your hand on the baby's head when it crowns.' The baby began to crown and I did as I was instructed, exchanging life force in the extraordinary circle of life.

Mother and daughter are an extraordinary team and this beautiful child, now all grown up, has been raised to know she can manifest anything and everything she needs. In fact they both do, traveling around the world manifesting what they need when they need it, living a life of joyous adventure.

Creative Alchemy for Art

Authentic art needs to be reinstated in our culture as an alchemical tool that is part and parcel of the healing process. Authentic art is pure expression of the inner landscape. The results may find footing within a particular artistic movement, for example the surrealists and the symbolists who endeavored to express ineffability. There is a need for art to return to a core purpose of alchemy, the transmutation of emotion by authentic creative expression.

From the point of view of Creative Alchemy, authentic art is a restoration of devotional practice and a psychological and emotional conduit. Taking the time to image what is stirring inside allows for a magical transformation and the formation of a crucible, the alchemist's tool, within which healing may take place. There are many noted artists who work in this way. Joseph Beuys, one of Germany's most well-known and provocative artists, used animal fat and felt in his extraordinary works, drawing from his life-saving experience of being wrapped in felt and animal fat by a Tatar tribesman after his plane crashed in Crimea. Anselm Kiefer, also German, expresses his interior reaction to the Holocaust, the burning of the Alexandrian Library and his own spiritual journey with majestic installations; huge paintings made from symbolic materials and sculptures that convey personal emotional charge and impact and are alchemical works of grandeur. Leonora Carrington was a British-born surrealist painter and novelist whose exquisite work provides stunning examples of authentic art: the visual examination of interior landscapes of the mind, emotions and soul. The exquisite artworks of the Australian Aboriginals are complex story and soul maps.

Authentic art can be powerful and thought-provoking. It can create great beauty or reflect on the horrors of life, as can work that has emerged from the commercialization of art and the branding of artists that began in the middle of the 20th century. The main difference is that one is an authentic and alchemical process and the other is a commentary. The two approaches can exist side by side.

The more energy we give to encouraging a more shamanic approach to art, and by that I mean the making visible of the invisible as opposed to cultural commentary, the more art can be restored to and function as a sacred alchemical tool. Part of this is insuring that all children are taught that we are fundamentally artists. Children need to be encouraged to have free emotional expression, a tenet of Creative Alchemy but also encouraged to make stand-alone works that share their interior life.

We are all artists creating continually, and if we all, from the smallest to the oldest, can be encouraged and allowed to express ourselves creatively, inner healing will become more automatic and with that we will have greater focus and ability to concentrate, greater alignment with purpose, relief from anxiety and a free flow of insight.

Creative Alchemy for Community and Mental Health

The cultivation of friendship and community is very important for our emotional and mental well-being. We are social beings. Getting involved in the support of some community events can help bring meaning to our lives as well.

For people who have become isolated, it's important to take steps. People working long hours are either too tired to socialize or feel the interaction with colleagues at work is enough. The truth is that meaningful interaction outside of work is irreplaceable. It can be very tempting to replace an intimate, caring social life with social media as well. I think social media is wonderful for connecting the world and sharing news, sometimes the good news and the bad news the traditional media doesn't print, but if we find ourselves being obsessive we have to take a step back.

Facilitating Creative Alchemy with high-risk individuals, I have found almost always that they are isolated and alone. They live alone, they don't have friendships and they are alienated from their families. It's easy for mental and emotional patterns to grow out of proportion when we are isolated. When we share our thoughts and feelings, we realize that we all share similar fears and hopes and dreams. People who suffer from the modern malaise of anxiety can learn that it is a common phenomenon and gain tools to work through it.

Drop-in centers perform a supportive function but organized circles provide the best results for those who are suffering. A safe place to share with like-minded people and those who have gained perspective can be life-changing. Many people having radical awakenings are being misdiagnosed as bipolar or schizophrenic and put on strong medication. This is across all ages. Mental health care needs to have an awakening. The curriculum of education for practitioners needs to include a broader understanding of what the expansion of consciousness looks like and how it can be supported.

The techniques of Creative Alchemy are very safe in this respect and help the release of unhealed emotion without the need to drug the individual. If mental health professionals were taught that unhealed, traumatized and suppressed emotion is the culprit and used these techniques for safe release, a different result would emerge for those afflicted. Mental health would be more accurately called emotional health. The emotions carry the power to affect the physical, mental and energy bodies. Our emotions draw on as much as 80 percent of our life force. When they are healed, free flowing, available for fuel and positive manifestation, releasing true vitality and personal power, all our earthly vehicles fall into alignment. With this newly gained harmony and inner peace we automatically begin to access the power of our Presence, the visionary capacity of our Higher Mind and the abundant supply of our Causal Body.

We often feel we have to hide these wounded parts of ourselves, but as I have discovered sharing these techniques with groups of teachers, business people and large groups of all ages and walks of life, releasing wounds in a loving and safe group actually strengthens bonds and creates enthusiasm, trust and support. Shame and guilt are also relieved as they are side effects of suppression. Creative Alchemy in groups has markedly helped people with depression, shame and hopelessness. Just knowing we are all in it together is a huge positive step.

There are many other things to do to help people out of isolation and expand their horizons. We can take a cooking course, or any number of courses. Or become a mentor as part of a mentorship programme for young adults. Join a local meditation group. Help out by sharing your gifts or giving talks at a local school. Get off the computer. Join a yoga class and ask someone you've been having chats with out for a coffee. Engage Visionary Imagination and Future Memory exercises to start shifting the imprint of your energy fields in a monthly group. Create a circle to discuss, share or further your passion. Create a Creative Alchemy study group!

The concept of the circle led by wisdom-keepers and elders is as old as humanity. Indigenous cultures say we are entering the time of the Sixth Sun, a return to the feminine, the magnetic power of the moon and the wisdom and community of the circle. This is good news.

Loneliness

A word about loneliness. Loneliness seems to be endemic on our planet. Since the Industrial Revolution 200 years ago, families have been far flung. The digital revolution of the 21st century has worsened this situation, as people do not congregate in shared workplaces as much as we once did. Friends who people grew up with are far away or forgotten, or friendships are never forged. Loneliness hits young and old and our suicide rates are climbing.

Most of the addictions we have we don't even call addictions. Nonetheless, these things that pull us away from activities of deeper fulfilment are trying to fill this hole of our loneliness. We are trapped behind the masks we wear, longing for authentic non-judgmental community.

The best cure for loneliness is connection to your Presence. Loss of connection to this part of us is a root cause of loneliness and enabling that reconnection is one of the main goals of Creative Alchemy. The more we flood ourselves with the energy of unconditional love that flows from the Presence to us, and the more we learn to also send it back and feel that exchange, the more our loneliness will heal. That empty hole inside which we try to fill up with endless distractions is a well of loneliness. Allow the Presence to fill it.

If loneliness is something you live with, it's important to make practical efforts to feel more fulfilled as well. We are social beings. Join a course, mentor a younger person, or volunteer once a week. As you become more deeply connected internally, your loneliness will start to lift and your outer world will reflect the energy of your new inner world. You will reach out to new friends more easily. Being with new people can be a useful step in dropping our old outworn identities. They have no expectations of who we are.

Any time this emotion hits, sit with it for a while. Don't distract yourself. Get to know it. Breathe into it. Get your journal out and write, to see what emerges from it. Once you have honored your loneliness, experienced it, find a way to express it. Dancing is a very good expression for sad feelings, as is drumming, which is why shamans use drums in their rituals. Rhythmic soft drumming moves us quickly into a Delta brainwave state and then into theta, where we are expanded into the comforts of unity consciousness.

Creative Alchemy for Prisons

Prisoners can be intelligent people who haven't been given the opportunity to develop the skills for a meaningful and abundant life. I became conscious of prisoners when my memoir, the story of my radical awakening and divine intervention, won an award in London from an organization that encouraged spiritual literature for prisoners.

Many of London's artists were at the award ceremony. Steven Berkoff, a highly noted British playwright, gave an address and stated that he was sure that if he hadn't been exposed to the arts he would have become a master criminal. That always stuck with me. How many people with great gifts are languishing in prisons because they never had a chance to express their gifts or because they were made to feel inferior by the system?

That book was distributed to prisons in the United States, the United Kingdom and New Zealand, and I received touching letters from inmates who had read it, conveying that they were able to see their beauty and worth for the first time. Many of the principles of Creative Alchemy have their origin in that book.

Our London theater occasionally brought theater into prisons. Inmates were strongly affected and empowered by this form of creative expression but it was not at all easy for the young directors who led the programmes. Had I known then what I know now, we would have included these steps of Creative Alchemy and woven them through the theatrical experience.

Anything that helps prisoners release trauma, gain self-worth, connect to their inner guidance and know their true essence is deeply valuable. Techniques such as Creative Alchemy, meditation and yoga begin to make lasting differences in lives. Testimonials from inmates regarding yoga in prison show relief from pain, anxiety, addictive behaviors, chronic pain and stress-related diseases. The system of EFT, Emotional Freedom Tapping, is also making great strides with PTSD (post-traumatic stress disorder).

The introduction into prisons of Vipassana, a ten-day silent meditation, has had profound effects. Ram Singh, who was home secretary of Rajasthan in India and witness to the early trials of Vipassana in prisons, states: '... *the success of the experiment heralds a new era of reform and rehabilitation for those who fall to crime. Vipassana provides an effective way to liberate them, not only from the life of crime but also from all suffering and misery.'* [200] There are moving stories of hardened criminals breaking down after a Vipassana retreat, opening their hearts and releasing all the stored pain they have carried their entire lives. Vipassana meditations are held in prisons in India, Israel, Mongolia, New Zealand, Taiwan, Thailand, UK, USA and Canada, as well as retreat centers worldwide.

Corrective training programmes are not going to dent individuals suffering from deep trauma, recurrent anxiety and chronic negative thinking and memory. Other methods such as those mentioned above are needed. If they could be helped to clear emotional trauma and taught how to connect to the inner guidance of their Higher Minds and Presence, their lives would change dramatically. It would also be extremely beneficial to have recently released prisoners be part of a weekly Creative Alchemy gathering and to foster the healing of emotion through systems such as Creative Alchemy while imprisoned. The pressure of the unhealed emotion must be released so that true and lasting change can occur.

Many have come from tragic and traumatic lives and are now expected to build new meaningful lives with very few tools or skills. Is it any wonder the unhealed emotion causing inner unrest drives them back to addictive and violent behavior when triggered? How easy would it be to address this?

This applies equally to children's homes where children are often feeling lost and abandoned and homes for the elderly.

Service

The true desire to serve is a consequence of illumination. We mess ourselves up with an imposed idea of what service is. It doesn't mean leaving the path of your gifts and talents and doing something foreign to your essence. It isn't a suppressive act and it isn't sacrificial, although there may be a bridging moment when we sacrifice a former identity.

True service is when we do what we absolutely love, and something shifts inside us so that we are doing what we love for the greater good as well as our own good. Self-service and world-service totally merge. There becomes no difference. True service is absolutely win-win.

You may experience a massive surge of power when this alignment occurs as it indicates that you have given yourself to your Presence. You are in alignment with the totality of the whole, and you are maintaining your uniqueness and your own sense of purpose, gifts and talents. Being of service with your gifts is a form of surrender that is ongoing. Your heart has grown and rather than diminishing your secret dreams, service has brought you more of everything you dream. When we are of service we no longer look *at* ourselves. We are too busy looking out *for* others.

Earth – the Great Mother

In the late 1960s British scientist and inventor Dr James Lovelock brought forth the Gaia Theory, named after the Greek primal Earth goddess, mother of all life. The theory proposed that Earth is a single living system, a self-regulating organism that works continually to maintain conditions for life. He was at first ridiculed by the scientific community. Formerly a scientist for NASA, Lovelock's theory gained support and was further developed by microbiologist Lynn Margulis and has since been used to explain how the oceans are kept in balance and why the atmosphere isn't mostly carbon dioxide. 'Life maintains conditions suitable for its own survival.'[201]

If the human has bones, Earth has stones; while the human has blood and lungs, so the Earth has oceans that rise and fall every six hours with the breathing of the world. As blood branches through us with veins, so water branches through Earth with rivers and tributaries.

Science has shown clearly how systems do intertwine and regulate one another and are deeply dependent on the healthy functioning of the whole. We can now measure global coherence and know how we ourselves affect Earth's geomagnetic field along with all life.

Divisions will continue to melt and more bridges will be built. In my adopted country, New Zealand, indigenous Māori culture holds an intricate, holistic and interconnected relationship with the natural world from which modern science is now learning. Their ancestral traditions influence their worldview of ecosystems as an interconnected interrelationship of all living things and create links between the state of ecosystems and cultural, physical and spiritual well-being. In general, the Māori outlook of ecological and human health takes in many more factors and in a more holistic and integrated way than the traditional Western view, taking into consideration life force, economic, social, cultural and environmental well-being and community when monitoring ecosystems. They include spirituality, emotions, mental health, cultural heritage, family and other factors when monitoring physical health. Doesn't it make perfect sense?

Warnings and prophesying from the Hopi of America to the Kogi of the Sierra Nevada de Santa Marta in Colombia, who see Earth as the Great Mother and themselves as spokespersons, have proved to be scientifically verifiable with time and catastrophes they predicted have occurred. Many of these could have been avoided if their warnings had been heeded. We, the younger brother, must come into alignment with the living entity who nourishes us, clothes our spirits and provides the materials for our manifestations. Our main tools in this endeavor are conscious consumption, respectful usage and infinite gratitude.

Science now confirms animals have a high level of sentience and emotional range and plants have their own sentience as well. The crystalline matrix of the planet can be measured to reveal a very high frequency. There are forms of intelligence other than our own. Consciousness permeates all. When you are manifesting, offering gratitude for all the elements that will clothe your dream is powerful.

Summing Up

Physicist, astronomer and mathematician Sir James Jeans said: 'The tendency of modern physics is to resolve the whole material universe into waves, and nothing but waves. These waves are of two kinds: bottled-up waves, which we call matter and unbottled waves, which we call radiation or light. If annihilation of matter occurs, the process is merely that of unbottling imprisoned wave-energy and setting it free to travel through space.' [202]

Let there be light [203] begins to sound less like poetry and more like the truth of the fabric of our reality. An intimation of the Big Bang by Old Testament authors? Or the first decree of the *I AM that I AM?* Either way, if we humans can directly influence this field of light with our own vision and emotion, where does this place us? Perhaps we are halfway between angel and animal, and the interface between the microcosm and the macrocosm of the universe.

Everything is light. This is the great mystery and the alchemical key. Brother Guy Consolmagno, research astronomer and President of the Vatican Observatory Foundation says, 'You never get to understand the universe. You just get used to it.' [204] Will we solve the mystery? Or embrace it?

I hope that in reading this book and applying the techniques of Creative Alchemy your perceptions and beliefs have begun to shift. The self is just the summation of programming and beliefs. Your programming can be changed. You can be whoever and whatever you wish in life. There is nothing in you that is fixed or static. The greater Self, the Presence, is All That Is.

I regard consciousness as fundamental.
I regard matter as derivative from consciousness.
We cannot get behind consciousness.
Everything that we talk about,
everything that we regard as existing,
postulates consciousness.[205]

– Max Planck,
theoretical physicist who originated quantum theory

Occasionally allow yourself to be nothing, nowhere and nobody. Enter the unified field and float. When you are really clear and able to become nothing, free of story, you can become everything and have access to everything.

You have created your current reality; you can create a new reality. It is possible to transcend our conflicting polarization and duality by embracing paradox as a concept and a practice. We create our reality from an infinite field of sentient energy. Science has revealed that only at the point of observation do the particles take a specific state and position. We are waveforms of light, projecting our lives from our pineal gland and heart, a dream on the screen of life. The moment we think that forces outside ourselves can affect our lives, we become victims. What we observe in the outside world is a direct result of how we think and feel. To create anew we must take our attention off the external and focus on the internal realm of the Presence. *The kingdom is within.*

You are majestic. You are powerful. You have the ability to heal yourself, regulate the systems of your body with your mind and emotions, live a life of meaning and purpose and create your own destiny. The whole universe deeply loves and supports you and is waiting patiently for you to take your place in the grand symphony. You *are* the universe.

8

MY STORY

There is the mud, and there is the lotus that grows out of the mud.
We need the mud in order to make the lotus.
— Thich Nhat Hanh

The seeds of this course came to life in 1995 after a massive personal expansion that lasted for two years.

I was a successful painter, theater director and filmmaker with an award-winning career in the arts. I ran a wonderful theater and pub in London with my second husband, its founder. We lived above it with my beloved daughter in a beautiful Victorian building. My husband had an incredibly keen eye for talent and we helped the early careers of many people who became household names. I also had a small practice helping primarily actors with the healing of emotion using techniques I had learned training in many modalities – the classic wounded healer. Life was glamorous, fun and abundant on the outside. But inside I was dealing with chronic anxiety, suppressed trauma, sleeplessness, low self-worth, paranoia and frequent panic attacks. I had all I needed for a great life, right? What could be wrong?

My childhood had been brutally challenging and I had been on a healing and intense spiritual journey since I was aged 12, seeking a way to relieve the inner torment and reunite with some lost part of myself. The part I was longing for I now know to be the Presence. I was spiritually aware, had prescient dreams and visions and longed to be lifted up and taken away by angels. I didn't understand why I was in this body and in this world.

During my childhood I experienced sexual abuse and chronic, extreme, physical violence. Later I experienced extreme psychological humiliation and mental and emotional abuse. My nervous system was hard wired to fluctuate between 'fight or flight' and 'paralysis.' But I was also able to enter bliss experiences from an early age and often experienced unity consciousness, my molecules seemingly flying away like a flock of birds and my consciousness merging with my surroundings, especially in nature.

My healing journey has provided first-hand information of what is effective and what is not. I really was a stranger in a strange land, a changeling with little to identify with in my surroundings. Sometimes my inner and outer life had no healthy 'gate.' The inner door was wide open, blending and blurring with the external reality. I could be in the depths of a despair so deep the thought of complete obliteration was a welcoming salve. I could just as easily be catapulted into a transcendent state where the division between heaven and Earth dissolved in an instant and the world became a visionary garden of translucent colors and sacred fire, a living wondrous dream. I was ungrounded and always skating on thin ice. Then came the 1960s. As perilous as the 1960s and '70s were for many, they were also a time of great awakening. A mass exodus from consensus reality and a breaking of the hypnotic hold of a limited worldview.

There was also something magical about my childhood. My mother was an artist and a poet with a wild unbridled soul, but she had no healing tools for her own inner trauma and projected her pain outwards with devastating consequences. My stepfather was a lead singer with a hauntingly beautiful voice, a musician and a bandleader who also expressed his unhealed emotions in very harmful ways. We traveled from place to place – frequently to Las Vegas, from one end of the USA to the other, with an ever-growing number of siblings on a mattress in the back of a pink Cadillac with fins. Life was rags to riches and hand to mouth and I began looking after my siblings when very young. Though I was the 'go to' person to care for the increasingly larger brood of my brother and sisters, there was also time to be forgotten and enter the world of the clouds, flowers, trees, other dimensions, and wonders of nature of which I felt very much a part. I had a bevy of imaginary friends I feel sure now were guides, who helped me get through. One gave me a mantra to say when I was eight or nine and which I did say frequently. It was decades later before I found out that *My strength is as the strength of ten because my heart is pure* was the mantra given to Sir Galahad in the story of Camelot in Alfred, Lord Tennyson's poem 'The Lady of Shallot.'

Living with my mother and stepfather, life was peripatetic, sometimes wondrous and often utterly terrifying. I lived in my mind, good for developing imagination. I lived above my body, seeking safety from the unexpected blows that could rain down without notice or the calculated ones where I was asked to choose the implement of abuse, and the paralyzing visitations in the night. Living outside the body, being *disassociated* as it's called in medical terms, is a good coping mechanism but not very good for being present in life! My mother taught me many things of value and painted, sang and wrote beautifully before life got the better of her. Much later I found out the trauma she had lived through, hidden behind the beautiful closed doors of my loving artistic grandparents. Trauma is passed down through generations. The abusers have almost always been abused. Time to break the chain of pain.

By contrast, my father's home was conservative and wealthy. My father was a brilliant man in continual conflict with the life he had created for himself. He was a beautiful soul who never felt he had fulfilled his true purpose. An artist or

My dear mother, I know what happened to you now. I can see it in your eyes. Your brother told my sister a while ago, a parting gift at the farewell door. I am going back in time to hold you close and tight and help you heal. Feel my arms around you. The chains of pain are broken and the river of life can flow freely again.

creative engineer in the body of a lawyer, the youngest Harvard graduate of his day. A vertical split in his psyche had created a bipolar pattern that afflicted him his entire life. He was a complicated man lost in his inner world who hugged like a great bear and had a warm deep voice but left my sister and me in penury and a culture of psychosis, abuse and violence. It was my stepmother who insisted we come to live with them.

This family didn't make art, they collected it. I went to live with them and my new siblings when things had clearly gone too far and they could not morally do otherwise. A 12-year-old wild child was not what my stepmother needed in her life. I was a crazy kid, a misfit, a rebel on a roller coaster of anxiety attacks, heightened visionary experience, depression to noisy life of the party – over the top, non-stop. She was a very powerful person and was an expert in prodding my low self-esteem and I think, looking back, was also afraid of me: my mysticism and otherworldliness was a huge contrast to her vital, present earthiness and materialism.

I came to be very fond of my stepmother later in life. She taught me an appreciation of beauty and culture. I came to understand the roots of her psychological persecution of me when I was young and their foundations in her own life.

The psychiatric treatment and medication began at 12 years of age. Lonely souls I thought at the time, some of whom opened up to me about their own lives. Rorschach tests (ink blot tests) showed me to be without healthy boundaries. I felt responsible for everything, connected deeply to everything, mystically converging with the unified field before I knew it had a name and plants, animals and open-hearted people. I suffered for the sufferers, felt their pain in my body. How many of us experience this lack of autonomy? There was no separation between me and the world. Perhaps the experience of oneness in an unhealed emotional body reaps adverse effects. The spiritual sages of the East recommended healing and overcoming patterns while incrementally building an internal structure to support this experience. I wonder how many individuals populate our mental health facilities who are having expanded experiences with no inner structure to support them? I also began meditating at 12. A friend taught me Nichiren Shōshū Buddhist meditation and I began devouring the work of Edgar Cayce.

My overreactions, intense overwhelming emotions and rebellion against the structures of society and all rigidity left a trail of chaos behind me. Looking back, I can see that another element was that I was awakening in a medicated and suppressed culture and my expanding energy field was causing uncomfortable catharsis in those around me. Those whose wings are clipped do not allow others to fly. It reminds them of their own buried pain and frustration. They sent me to a boarding school for misfit children where I found many other like-minded souls, some of whom are friends to this day. Later, I traveled. My father and stepmother were more than happy for a relief.

I sought out sages and shamans and sat at the feet of spiritual masters and wisdom teachers from Native American, Tibetan, Russian, European, Balinese, Chinese, Ghanaian, Indian, Peruvian and Mexican traditions, learning and sincerely applying esoteric systems of meditation and healing. I trained in many modalities including Reiki, Rebirthing, Magnified Healing, Craniosacral therapy, Energy Healing, Genome Healing, Psychology, Inner Mystery School, Ascended Master Teaching and Trance Dance, while experiencing many others.

The 1960s was the era that bridged ancient metaphysical wisdom and the new science of quantum theory, creating a cohesive worldview that has much to do with the ongoing awakening to unity consciousness we are currently experiencing. It set the stage for the shift of the ages – our metamorphosis from Homo sapiens to Homo illuminatus. Quantum theorists such as Fritjof Capra emerged with astounding revelations captured in *The Tao of Physics*. Abraham

Maslow created the 'pyramid of need' concept and the Esalen Institute was founded in Big Sur, California, playing a key role in the human potential movement. The study of the expansion of consciousness and peak experiences as a necessary part of the human experience available to all, holistic thinking and the 'East meets West' revolution of thought entered the mainstream. The core truths of world religions and obscure mystical and hermetic texts were compared and found to be more similar than different. It was a time of unparalleled seeking.

An entire generation rose up against war, the birth control pill was co-invented by chemist Carl Djerassi (who later wrote plays we presented at the theater I co-ran), and scientist Albert Hoffman fathered the psychedelic movement. Our default-mode networks were offline, and we were expanding. The status quo was taking a beating but we were to later learn that 'tuning in, turning on and dropping out' doesn't change the world. I learned so much about the human mind and spirit and myself but did not manage to shift the deep core wound driving me internally.

I threw myself at life and studied continually, attended art college, created art, painted feverishly, made huge found-object sculptures and shadow boxes inspired by the artist Joseph Cornell, wrote long free-associative prose and poems and got into lots of trouble. I adored art, its mystery and ability to hold something of the psyche and soul of its creator and also the culture it reflected – its alchemical nature when it is authentic

Older now and free of medication (but not yet counterculture medicine) and the accelerating properties of teenage hormones, I still experienced periods of ecstatic joy when the beauty of life seared me. I could *feel* everything. There was no filter. Creating helped me, keeping me sane and helping me deal with painful mood swings that could bloom into depressions falling like a lead blanket. One psychiatrist I went to in my twenties who had lived through the Holocaust said the continual unexpected violence of my childhood was similar to what she had experienced in the camps. *'You're doing really well,'* she had said, *'you haven't killed yourself.'* Everything is relative. In fact, I had contemplated it seriously, but something stopped me. I still have a scar on my left wrist. I believe systems like Creative Alchemy have the potential to greatly assist young people contemplating committing suicide worldwide at an unprecedented level. I was almost one of them. Having experienced so much myself has left me with enormous compassion for these young people. How can we help them? Can we get Creative Alchemy into every school so children can learn emotional literacy, the power of Living Vision and the truth of their majestic Presence?

Flash forward to the mid-1990s. Still not entirely embodied nor completely connected to my Source despite continual seeking, I used the will of my mental body rather than my Presence and physical vitality to push large film, theater and art projects into being. I was exhausted. One day I finished reading a book written by someone who has made a very positive contribution to helping the world, Patricia Cota-Robles, and I asked plaintively: *When will I make a contribution that is meaningful? When will I be Okay?* I heard an answer as clear as day: *What makes you think you will not burn as bright?*

I think, in that moment, somehow my radio tuned into more subtle sound waves, or maybe I was just so desperate they took pity on me! I asked for community and to be 'seen.' An hour later there was a knock on my door and Christina Hagman, whom I had met on a healing course, presented herself. That began a two-year personal course in miracles. We started an earnest metaphysical exploration of the subtle realms, guided by teachers who had once walked the Earth, who had graduated from Earth university with flying colors and now were dedicated to helping the rest of us from the ascended realms along with other cosmic beings. Many things happened during the two years that followed. It was a time of continual inspiration, mind shifts, transcendent experience and deep healing. I was guided to keep a journal. Those journals became a memoir, a 'warts and all' no BS depiction of the journey to awakening. I was able to notate huge expansions as they occurred, and also the contractions that can follow in the early stages, which can leave us bewildered.

The masters who guided us gave us many healing modalities together and separately. We egged each other on through full-blown cathartic healings, taking turns facilitating each other, gradually lessening the internal pressure. We were drunk on light. Everything started shifting. But with it came fear, as surrender of the outer reality is necessary for total transformation and that takes time. The 'ego' is afraid it's going to die and will do anything to prevent it. Radical Acceptance and love are the key. Once we begin this alchemical journey and hitch our wagon (bodies) to a star (our Presence), there is no going back.

Receiving the basics of what would become Creative Alchemy was the most important inspiration in those two years, and applying a heuristic (learning from your journey rather than the journeys of others) testing process over several years insured that the results of the protocol were profoundly understood, later confirmed by facilitating others. The profound healing did not happen overnight. We were diligent. We followed the directives every day like wilted flowers opening to finally receive nourishing rain. And then we began to bloom. It was working. The Accelerated Body Communication you found in Part Two is a modality Christina and I received together. We facilitated each other through an entire summer with that one exercise, our cries drowned in the traffic of the London streets above the busy pub and theater as the residues of our traumatic childhoods were released, pushing each other through resistance.

Christina received systems she facilitated with others when she returned to her home country, Sweden, and became an expert on the 'game of abandonment,' which holds humanity in crisis. She developed a process to help connect people to their Presence. She went into the prison system to work with abusers, understanding 90 percent of all abusers have been abused themselves and the punishment of incarceration must be accompanied by therapy if we are going to heal as a culture. I carried on in England, developing this system and then teaching it, including a 12-week trial with suicidal patients in London. Now, all these years later, I have been invited to facilitate this process in four countries for countless people of all ages and walks of life, often very mysteriously and with no promotion, continually refining it. I have always thought of it as my service. The force that delivered it seems to be also in charge of it. It became time to write it down, share the system with a larger audience and pass it on by training teachers, which I have begun. Time to properly pay it forward.

What is my life now? I can hardly believe it. It didn't happen overnight, and I chose some huge projects to manifest early on including a feature film, learning painfully how crucial healing emotion is before tackling large manifestations. But now my life is living me.

What does that mean? I'm off the wheel of time for the most part. I am on I AM Presence time. I have loving friends and family. I love my work and I am abundant. The perfect amount seems to come when I need it, as long as the purpose is in perfect alignment. I trust pretty deeply and consistently. I have day-to-day things to attend to, but they don't stress me as they used to. I can still get thrown but it's rare and then I practice the steps of Creative Alchemy in a sort of shorthand after all these years. Sometimes I just stop and truly experience the emotion and breathe, that's enough. Sometimes I just accept I am grumpy, if that's the case. Sometimes persistent mood shifts are an indication we need to radically change something in our lives and it is not internal at all. There are still patterns to clear but the awareness is greater than the patterns.

I am consistently motivated, disciplined and empowered to do the work I feel is my true purpose and it is a joy. This includes producing creative works that are underpinned with the understandings of Creative Alchemy – painting, novels, screenplays and so on and of course, sharing the system of Creative Alchemy with as many people as possible. There are long periods where I feel quite neutral but very dedicated and motivated; and then there are precious states of ecstatic joy as more of the Presence flows through me. Is this how we could all be living? Yes!

Sometimes there is God, so quickly!
– Tennessee Williams, *A Streetcar Named Desire*

I was a bloomin' hard case. Imagine how much easier it's going to be for you. If you've also been through a lot, have hope. It's going to turn around now.

The more I practice the Presence, keeping my attention on my decrees and the force of my life and also using the Law of Forgiveness and the Violet Ray to cleanse cause, core, record, effect and memory of that which interferes with my total connection to my Presence, the more aligned I am. Clairvoyants can always tell I am a person who decrees the Violet Flame because my auric field is primarily violet! I also spend time every day sending the flame to every energy field, plant, animal, elemental, human and angel in our world and in the Universe. I send it to the sun and to the sea and the Earth and the galaxy. Some other part of me seems to be doing it spontaneously. You can also send it specifically to areas in need or children in need and so on. This is crucial. The more we give, the more we receive. I feel driven by Light, the alchemical key. I think I have crossed the bridge of surrender and the force of Light is literally transforming me. This is the promise of Creative Alchemy should you wish to go 'all the way.'

Paramahansa Yogananda states that if we remember that: *God has become me'* [I AM my I AM Presence] *and accordingly act divinely* [in alignment] *you will be liberated from the imperfections of your body and mind.* Creative Alchemy has combined powerful emotional healing, the activation of the imagination to Living Vision and the all-important final piece, inner oneness with the Presence, whose Light and Fire is all-transforming.

Great care has been taken to make the material accessible to modern minds backed by discoveries in our sciences. Dogma and blind belief have no part in it. It is experiential and observational, alive and evolving. Thanks to the discoveries of quantum mechanics, the science of consciousness and the science of energy are merging. I believe that it is science that will ultimately prove the reality of the Presence, not theology.

I am finally fully in my body. I credit New Zealand, my family and my friends here for that. People live close to nature here and they are very grounded on the whole, very chilled and not overly competitive but quietly committed to excellence. Being embodied makes me feel safe. I can access my body cognition now and therefore my discernment is so much higher. How we relate to our bodies is very much how we relate to our world. I don't have many thoughts in my head. No negative chatter, which is astounding. I had such inner torment earlier in my life. It has given me huge compassion for people who suffer this way and huge delight when Creative Alchemy relieves this for people.

I still need to be vigilant and I am careful with myself. I've given myself permission to be who and what I am. I don't force myself and I no longer choose pleasing others over listening to my body, as hard as that can be for myself and loved ones. We simply can't anymore. It's about discovering who we really are, our true essence and what suits us. The more of us who can be present, the more we can be of service and of true value to others and ourselves. Self-care is crucial. When we are full our cup runs over and we give without depletion.

I couldn't have chosen a better childhood to compel me to figure it all out so I could go on to create the system of Creative Alchemy and help you now. Hopefully your journey of emotional healing is not nearly so severe. I sincerely believe we all have some trauma to clear on the great journey of 'Know Thyself' and when we are clear, the microcosm and the macrocosm reveal themselves to be one. When we clear, we do so for our lineage as well, and it's important to state this intention.

I have a strong meditation practice using the Kriya Yoga breathing technique taught by Paramahansa Yogananda and the foundation that carries on his work, Self-Realization Fellowship, as mentioned earlier.[206] This has helped me with inner alignment and embodiment and I feel I am under the force field of the great sages of that lineage whose Presences are often very audible and palpable. I have a strong prayer practice as well. My particular areas of prayer are for children of abuse and human trafficking. Millions of others join me in this and there seems to be increasing highlight on this tragedy and more action being taken. I am blessed to have two highly advanced spiritual mentors, one for 25 years, both to whom I am eternally grateful. It's a never-ending journey!

The silver lining to your healing journey is that it led you to the path of the Great Work, the alchemy of Spirit by which we turn the lead of the psyche to radiant gold and become all we can be, all we are destined to be – illumined beings of limitless potential. I have learned first-hand of the ability of the human Spirit to triumph over adversity again and again. We are all phoenixes rising from the fire.

It's time to love what you do and do what you love. Get your inspiration from your Presence and carry out the activity, all the steps. Do whatever you are guided to do, and don't stress about what happens next. It will be what it's supposed to be. If you need to reverse-engineer and readjust, you do that. Suddenly the key fits the lock and the door opens, finally never to be shut again. The outer reality can no longer disturb you. *I AM the open door no man can shut!*

This is what I sincerely want for you. My Future Memory for this book is you fulfilled, you joyful, you with meaningful purpose, abundance, health and prosperity. You with loving people all around you who see your true essence and your true beauty and support you on your journey as you support them. Please join me on the path of the Great Work and sincerely apply the steps of Creative Alchemy. I'm holding the space for you.

I AM the full activity of the Presence in action – all else has no power to move me – nothing else shall ever pass into my world, mind and affairs, surrounded right now as I AM by my Presence's canopy of sacred fire! I AM the Cosmic Victory made manifest Now and Forever in all timelines and dimensions for the highest good of all. And so it is.

I leave you with two powerful thoughts. One from pioneering physicist Sir James Jeans: 'The stream of knowledge is heading toward a non-mechanical reality; the universe begins to look more like a great thought than like a great machine. Mind no longer appears to be an accidental intruder into the realm of matter, we ought rather hail it as the creator and governor of the realm of matter. Get over it, and accept the inarguable conclusion. The universe is immaterial – mental and spiritual.'[207] And the other from professor of physics and astronomy at Johns Hopkins University, Richard Conn Henry: 'A fundamental conclusion of the new physics also acknowledges that the observer creates the reality. As observers, we are personally involved with the creation of our own reality. Physicists are being forced to admit that the universe is a mental construction.'[208]

Go forth and manifest a beautiful reality and let's co-create the new world together.

ACKNOWLEDGMENTS

Gratitude must first go to the guides and masters of all realms who have helped me on the path and continue to as I integrate more and more of my Presence and Higher Mind into my earthly bodies, especially the great Masters of Alchemy. I must also thank Christina Hagman with all my heart and soul. In this material, she has been the magnetic South of my North. Whatever hopeful merit Creative Alchemy may bring to the world, our time together in 1995 and our connection and communication since then has been foundational.

I want to thank my beloved daughter, Katherine, and my son-in-law, Fraser, for their continual nurturing support since I arrived in New Zealand, Land of the First Light, which hugely helped me find the inner stillness to write this book and progress along the path of service. I thank my granddaughter Arielle for often seeing what I see and surprising me with what she sees that I don't and for being a continual source of huge delight. Thanks also for the deep pleasure of having my joyous grandson, Corin, and my new little granddaughter Casey in my life, and the continual nourishing expansion of the heart my family provides.

I want to thank my New Zealand friends who have shared the journey with me. You have all been incredible, each so unique and dear. I thank you. I want to thank New Zealand herself. Beautiful Aotearoa, your sacred lands and mermaid-filled seas, your strong cleansing winds and your chilled attitude – all have been so healing and restorative.

Special thanks to Bob Davidow, who was the wind beneath my sails during the writing of this book, and to Viculp Lal for his deep insights on Indian culture.

Thanks goes to Stephanie Gunning for her first pass. She really helped me raise my game. Dr Rebecca Weiss-Vlasic, a PsyD in clinical psychology, gave me extensive notes for which I am deeply grateful. A lot of this material is her territory, especially the interface between psychology and biology, and she was an incisive force as an editor. She also happens to be my beloved sister. So, huge gratitude for a gracious and generous act of sisterly love. I also deeply thank Gabrielle Meech, former human rights lawyer and teacher, who gave the book a third edit, concentrating on metaphysical content. I also want to thank Shirley Walden for her thorough proofing and the lovely friends who have encouraged me so much along the way and really believed in this book. Thanks to my two stellar Wellington local cafés for providing sanctuary during the writing of this book, Archimboldi and Park Kitchen. Thank you to all at

Left Bank Parking Design for their work in bringing this book to publication, especially Paul Stewart, whose stellar layout and design helped win the Ashton Wylie Award for this book.

To my late husband, Dan: You taught me so much in our 21 years together, especially focus, dedication, discipline and the tempering by fire of the 'hero's journey' of intimate relationship. Matthew, you co-founded the feast of the amazing family we share and have been my brother since age 14. Dad and Mom, wherever you are, whatever stars you are dancing on, you were two extraordinary, unique characters. It was never dull! Thank you for being my doorway to this life, the lessons learned, the magic you imbued. And for having me!

NOTES AND REFERENCES

Due to the dynamic and ever-changing nature of the internet, some of these links may no longer be active. Apologies in advance if you find that to be the case.

Introduction

1 HeartMath Institute. Global Coherence Research, The Science of Interconnectivity. https://www.heartmath.org/research/global-coherence/

2 Max Planck. *Das Wesen der Materie* (*The Nature of Matter*), a 1944 speech in Florence, Italy. Archiv zur Geschichte der Max-Planck-Gessellschaft, Abt. Va, Rep. 11 Planck, Nr. 1797. The full quote: 'As a man who has devoted his whole life to the most clearheaded science, to the study of matter, I can tell you as a result of my research about atoms this much: There is no matter as such! All matter originates and exists only by virtue of a force that brings the particles of an atom to vibration and holds the most minute solar system of the atom together. We must assume behind this force the existence of a conscious and Intelligent Spirit. This Spirit is the matrix of all matter.'

Part One

3 This is paraphrased from *Bridge of Freedom*, Book 1. Maha Chohan, p.8.

Our Quantum Generator – The Power of Emotion

4 Dr Joe Vitale and Ihaleakala Hew Len PhD. *Zero Limits*, p.31.

5 Dr Sarah McKay, neuroscientist and author of Your Brain Health blog.

6 Manuela Lenzen. 'Feeling Our Emotions', *Scientific American*. www.scientificamerican.com/article/feeling-our-emotions/

7 Ibid.

8 Jonathan R Zadra and Gerald L Clore. 'Emotion and Perception: The Role of Affective Information'. *PMC*, US National Library of Medicine. www.ncbi.nlm.nih.gov/pmc/articles/PMC3203022/

9 Chai M Tyng, Hafeez U Amin and Mohamad N M Saad. 'The Influences of Emotion on Learning and Memory', *Frontiers in Psychology*. www.ncbi.nlm.nih.gov/pmc/articles/PMC5573739/

10 Julie Hani, RN, BSN, BA, CDE. 'The Neuroscience of Behavior Change' (8 August 2017). www.healthtransformer.co/the-neuroscience-of-behavior-change-bcb567fa83c1

11 Jill L Kays PhD, Robin A Hurley MD, and Katherine H Taber PhD. 'The Dynamic Brain: Neuroplasticity and Mental Health', *The Journal of Neuropsychiatry and Clinical Neuroscience*. www.neuro.psychiatryonline.org/doi/full/10.1176/appi.neuropsych.12050109Psychiatry online

12 Rollin McCraty, Mike Atkinson, and Dana Tomasino. 'Modulation of DNA Conformation by Heart-Focused Intention'. *HeartMath Institute* (2003), pp.1–6. www.aipro.info/drive/File/224.pdf.

13 Arjun Walia. 'How Earth's Magnetic Fields and "Human Aura" Carry Biological

Information Connecting All Living Systems, Collective Evolution' (14 March 2018) www.collective-evolution.com/2018/03/14/how-earths-magnetic-fields-human-aura-carry-biological-information-connecting-all-living-systems/

14 Candace B Pert PhD. *Your Body is Your Subconscious Mind: Mind-Body Medicine Becomes the Science of Psychoneuroimmunology.* www.healingcancer.info/ebook/candace-pert

15 Candace B Pert PhD, et al. 'Neuropeptides and Their Receptors: A Psychosomatic Network', *Journal of Immunology*, Vol. 135, Supplement 2 (August 1985), pp. 820–826. www.ncbi.nlm.nih.gov/pubmed/2989371.

16 Candace B Pert PhD, et al. 'The Psychosomatic Network: Foundations of Mind-Body Medicine', *Alternative Therapies in Health and Medicine*, Vol. 4, No. 4 (July 1998), pp.30–41. www.ncbi.nlm.nih.gov/pubmed/9656499.

17 Candace B Pert PhD. *The Wisdom of Receptors: Neuropeptides, Emotions and Body-Mind*, pp. 1, 5–6. candacepert.com/wp-content/.../Advances-v8-1988-Wisdom-of-the-Receptors1.pdf

18 Kristin W Samuelson PhD. *Post-traumatic Stress Disorder and Declarative Memory Functioning: A Review.* www.ncbi.nlm.nih.gov/pmc/articles/PMC3182004/

19 Kristalyn Salters-Pednault PhD. 'Suppressing Emotion and Borderline Personality Disorder', *Very Well Mind.* www.verywellmind.com/suppressing-emotions-425391

20 Amon Buchbinger. 'Out of Our Heads: Philip Shepherd on the Brain in Our Belly', *The Sun*, interview April 2013. www.thesunmagazine.org/issues/448/out-of-our-heads

21 Plato. *Timaeus*, trans. Benjamin Jowett. www.classics.mit.edu/Plato/timaeus.html

22 Ibid.

23 Richard Dawkins. *The Selfish Gene* (1976) p. 2.

24 Simo Knuuttila. 'Emotions from Plato to the Renaissance', *Sourcebook for the History of the Philosophy of Mind.* Spring Link (28 July 2013), pp. 463–497. www.link.springer.com/chapter/10.1007/978-94-007-6967-0_29.

25 Jini Reddy. 'What Colombia's Kogi Can Teach Us About the Environment' (29 Oct. 2013). www.theguardian.com/sustainable-business/colombia-kogi-environment-destruction

26 Rollin McCraty. 'Exploring the Role of the Heart in Human Performance', *Science of the Heart* (1993–2001). www.heartmath.org/resources/downloads/science-of-the-heart/

27 'The Heart as a Hormonal Gland', Chapter 1, ibid.

28 Rollin McCraty PhD and Doc Childre. *Coherence: Bridging Personal, Social and Global Health* (2010). www.heartmath.org/research/research-library/basic/coherence-bridging-personal-social-and-global-health/

29 Rollin McCraty, Mike Atkinson, and Dana Tomasino. 'Modulation of DNA Conformation by Heart-Focused Intention', *HeartMath Institute* (2003), pp.1–6. www.aipro.info/drive/File/224.pdf.

30 'You Can Change Your DNA', *HeartMath Institute* (July 14, 2011). www.heartmath.org/articles-of-the-heart/personal-development/you-can-change-your-dna/

31 Anontio Damasio, 'Why Your Biology Runs on Feelings', *Nautilus*, (18 Jan. 2018). http://nautil.us/issue/56/perspective/why-your-biology-runs-on-feelings

32 'Neuro-cardiology: The Brain on the Heart', *Science of the Heart*, Vol. 1, Chapter 1, p. 4. www.heartmath.org/resources/downloads/science-of-the-heart/

33 Justin Sonnenburg and Erica Sonnenburg. 'Gut Feelings – the "Second Brain in Our Gastrointestinal Systems"', *Scientific American.* www.scientificamerican.com/article/

gut-feelings-the-second-brain-in-our-gastrointestinal-systems-excerpt/

34   Gregg Braden. 'The Brain in Your Chest: Science-Backed Techniques for Tapping into Your Heart Intelligence', *Conscious Lifestyle*. www.consciouslifestylemag.com/heart-intelligence/

35   Clifford N Lazarus PhD. 'Does Consciousness Exist Outside of the Brain?' *Psychology Today*, www.psychologytoday.com (26 July 2019)

36   Dr Bruce Lipton, 'Your Body is An Illusion', www.youtube.com/watch?v=uCIgxYuNGuo

37   Antonio R Damasio. 'Feeling Our Emotions', *Scientific American Mind*. www.scientificamerican.com/article/feeling-our-emotions/

38   Personal conversation with international PEM instructor Sarah Victoria, PEM Workshop, Toi Whakaari, NZ Drama School, Wellington, NZ.

39   Bessel van der Kolk. *The Body Keeps the Score: Brain, Mind, and Body in the Healing of Trauma* (New York, Viking, 2014), p. 43.

40   Matt Danzico. 'Brains of Buddhist Monks Scanned in Meditation Study' BBC News (24 April 2011). www.bbc.com/news/world-us-canada-12661646

41   Alex Huth, et al. 'Decoding the Semantic Content of Natural Movies from Human Brain Activity', *Frontiers in Systems Neuroscience* (7 Oct. 2016). www.frontiersin.org/articles/10.3389/fnsys.2016.00081/full. Also: https://www.theguardian.com/science/2016/apr/27/brain-atlas-showing-how-words-are-organized-neuroscience)

42   The Gospel of Thomas trans. Thomas O. Lambdin. Gnostic Society Library/Nag Hammadi Library (accessed 17 June 2018). gnosis.org/naghamm/gthlamb.html.

Our Quantum Navigator – The Power of the Body

43   Walt Whitman. *I Sing the Body Electric*, Verse 8.

44   Ed Decker. 'Your Heart and Stomach May Be Smarter Than You Think', *Rewire Me* (3 Dec. 2013). www.rewireme.com/brain-insight/your-heart-and-stomach-may-be-smarter-than-you-think.

45   Richard Gray. 'Phobias May be Memories Passed Down in Genes from Ancestors', *Science Correspondent*. www.telegraph.co.uk/news/science/science-news/10486479/Phobias-may-be -memories-passed-down-in-genes-from-ancestors.html

46   Ibid.

47   Mordehai Heiblum et al. Weizmann Institute of Science. www.sciencedaily.com/releases/1998/02/980227055013.htm.

48   Alex Vikoulov. 'The Unified Field and the Quantum Nature of Consciousness', *Ecstadelic* www.ecstadelic.net/ecstadelic/the-unified-field-and-the-quantum-nature-of-consciousness.

49   Max Planck. *Das Wesen der Materie (The Nature of Matter)*, a 1944 speech in Florence, Italy. Archiv zur Geschichte der Max-Planck-Gesellschaft, Abt. Va, Rep. 11 Planck, Nr. 1797.

50   Walt Whitman. 'I Sing the Body Electric', Verse 9.

51   Statement of 1963, as cited by Walter J Moore. *Schrödinger: Life and Thought* (1992), p. 1

52   Scientific Foundation of the HeartMath System. *Two Way Communication*. HeartMath Institute. www.heartmath.org/science/

53   'The Energetic Heart is Unfolding', *Science of the Heart*. HeartMath Institute (22 July 2010). https://www.heartmath.org/articles-of-the-heart/science-of-the-heart/the-energetic-heart-is-unfolding/

54 Rollin McCraty PhD. Energetic Communication, *Science of the Heart: Exploring the Role of the Heart in Human Performance,* Vol. 2, pp.36–44, Boulder Creek, CA, HeartMath Institute (2015). https://www.heartmath.org/research/science-of-the-heart.

55 Lea Winerman. *The Mind's Mirror.* https://www.apa.org/monitor/oct05/mirror.aspx

56 Sadie F Dingfelder. *Autism's Smoking Gun?* https://www.apa.org/monitor/oct05/autism.aspx

57 Jessica M Yano, et al. Indigenous Bacteria from the Gut Microbiota Regulate Host Serotonin Biosynthesis, *Cell,* Vol. 161, No. 2 (9 April 2015), pp.264–76.

58 Adam Hadhazy. Think Twice: How the Gut's 'Second Brain' Influences Mood and Well-being, *Scientific American* (12 Feb. 2010). www.scientificamerican.com/article/gut-second-brain.

59 Ed Decker. Your Heart and Stomach May Be Smarter Than You Think, *Rewire Me* (3 Dec. 2013). www.rewireme.com/brain-insight/your-heart-and-stomach-may-be-smarter-than-you-think.

60 Sun Tui's website: www.IFEELcenter.com

61 Chakras, Karma and The Inner Solar System, *Yoga Yukta Life,* (21 Jan. 2016). www.yogayuktalife.com/articles/2016/6/2/chakras-karma-the-inner-solar-system

62 http://candacepert.com/achievements/

63 *Things that Make You Go Om,* Tag Archives. https://sevenintentions.wordpress.com/tag/jung/,

64 Renee Weber. *Dialogues with Scientists and Sages: Search for Unity in Science and Mysticism* (1986) p.44. The full quote by David Bohm: 'Mass phenomenon of connecting light rays which go back and forth, sort of freezing them into a pattern. So, matter, as it were, is condensed or frozen light.'

65 Paramahansa Yogananda, astral sounds, https://www.yogananda.com.au/g/g_astral_sounds.html

66 Patricia Cota-Robles. *Activating Our 5th Dimensional Solar Spine and Twelve Solar Chakras.* https://eraofpeace.org/pages/premium-content/?mc_cid=44fbfe45e2&mc_eid=a1e56a38a0#media-popup

67 Sigrid Breit, et al. 'Vagus Nerve as Modulator of the Brain-Gut Axis in Psychiatric and Inflammatory Disorders', *Frontiers in Psychiatry* (13 March 2018). www.frontiersin.org/articles/10.3389/fpsyt.2018.00044/full.

68 Paramahansa Yogananda. *Autobiography of a Yogi,* Chapter 26, pp. 269–272, Kindle.

69 Mahasamadhi. www.yogapedia.com/definition/5825/mahasamadhi

70 Paramahansa Yogananda. *Autobiography of a Yogi,* p. 548, Kindle.

Our Quantum Conductor – The Power of Sound and Energy

71 Plutarch. *Convivialium disputationum,* liber 8, 2.

72 Robert Lamb. 'How are Fibonacci Numbers Expressed in Nature?' HowStuffWorks. https://science.howstuffworks.com/math-concepts/fibonacci-nature1.htm

73 Albert Van Helden. Galileo, *Encyclopaedia Britannica.* https://www.britannica.com/biography/Galileo-Galilei/

74 Paul Halpern. Quantum Harmonies: Modern Physics and Music, *Nature of Reality,* Public Broadcasting System (10 Sept. 2014). http://www.pbs.org/wgbh/nova/blogs/physics/2014/09/quantum-harmonies-modern-physics-and-music

75 Max Tegmark. 'Is the Universe Made of Math?' *Scientific American,* (10 Jan. 2014).

https://www.scientificamerican.com/article/is-the-universe-made-of-math-excerpt/

76 Paul Halpern. Quantum Harmonies: Modern Physics and Music, *The Nature of Reality*, (10 Sept. 2014). http://www.pbs.org/wgbh/nova/blogs/physics/2014/09/quantum-harmonies-modern-physics-and-music/

77 Louis de Broglie. *The Information Philosopher*. http://www.informationphilosopher.com/solutions/scientists/de_broglie/

78 Dr Fritz-Albert-Popp. *Biontology Arizona*. https://www.biontologyarizona.com/dr-fritz-albert-popp.

79 Eugene T Gendlin. The Concept of Congruence Reformulated in Terms of Experiencing, *Counselling Center Discussion Papers*, Vol. 5, No. 12 (1959) p.1. University of Chicago Library.

80 Rollin McCraty. *Science of the Heart: Exploring the Role of the Heart in Human Performance*, Vol. 2, Chapter 4. HeartMath Institute (2015). https://www.heartmath.org/research/science-of-the-heart.

81 Hans Jenny. Cymatics: The Sculpture of Vibrations 1: Patterns of a World Permeated by Rhythm, p.6. *Unesco Courier,* Paris, France (Dec.1969). http://unesdoc.unesco.org/images/0007/000782/078290eo.pdf.

82 *What Is the Mind?* Quantum Training Institute. www.quantumtraininginstitute.com/what-is-the-mind.

83 John 1:1, *World English Bible.* (Public Domain)

84 John 1:14 *KJV.* (Public Domain)

85 Brendan D. Murphy. *Sound into Form: Cymatic Insights and the Sri Yantra.* www.globalfreedommovement.org/sound-into-form-cymatics-insights-and-the-sri-yantra/

86 Risha Joshi. 'The Pineal Gland and the Symbol of Manifestation: The Sri Yantra', *Power of Thought Meditation Club*, (31 August 2015). http://powerthoughtsmeditationclub.com/the-pineal-gland-symbol-of-manifestation-the-sri-yantra/

87 David Furlong. The Osirian and the Flower of Life, *Egyptian Tour* (accessed 26 July 2018). http://www.davidfurlong.co.uk/egypttour_osirion.html.

88 Masaru Emoto. What Is the Photograph of Frozen Water Crystals? *and* Water Crystal Photo Gallery. http://www.masaru-emoto.net/english/water-crystal.html.

89 Bernd Kröplin and Regine C Henschel. *Water and Its Memory: New Astonishing Insights in Water Research,* trans. Ulrich Magin and Regine C. Henschel. Stuttgart, Germany, GutesBuch Verlag (2017). Kindle, Loc. 72, 771, 786, 819.

Our Quantum Manifestor – The Power of Imagination

90 Sara Warber MD. *Walking off Depression and Beating Stress Outdoors? Nature Groups Linked to Improved Mental Health.* Study conducted by the University of Michigan, with partners from De Montfort University, James Hutton Institute, and Edge Hill University in the United Kingdom (23 Oct. 2014). https://medicine.umich.edu/dept/family-medicine/news/archive/201410/walking-depression-beating-stress-outdoors-nature-group-walks-linked-improved-mental-health

91 Neil Hermann. What Is the Function of the Various Brainwaves? Scientific American (accessed 28 July 2018). https://www.scientificamerican.com/article/what-is-the-function-of-t-1997-12-22.

92 *What Are Brainwaves?* Brain Works, Methods, Neuro-feedback. https://brainworksneurotherapy.com/what-are-brainwaves.

93 James A Blumenthal PhD, Patrick J. Smith PhD, and Benson M. Hoffman PhD.

Is Exercise a Viable Treatment for Depression? https://www.ncbi.nlm.nih.gov/pmc/articles/PMC3674785

94 Namdev. *The Philosophy of the Divine Name.*

95 Olivia Solon. Is Our World a Simulation? Why Some Scientists Say It's More Likely Than Not. *Guardian* (11 Oct. 2016). https://www.theguardian.com/technology/2016/oct/11/simulated-world-elon-musk-the-matrix/ Also: Nick Bostrom. Are You Living in a Computer Simulation? *Philosophical Quarterly,* Vol. 53, No. 211 (2003), pp.243–255. https://www.simulation-argument.com/simulation.pdf.

96 Brian Whitworth. *The Physical World as a Virtual Reality.* Cornell University Library. www.arxiv.org/abs/0801.0337.

97 Elizabeth Barrett Browning. *Aurora Leigh,* (1856). Book seven, lines 812–826. Read poem in full here: www.digital.library.upenn.edu/women/barrett/aurora/aurora.html

98 Samuel Taylor Coleridge. *Biographia Literaria,* (1817). Chapter 13, p.378

99 Ibid.

100 Physicists Continue Work to Abolish Time as Fourth Dimension of Space, *Phys. org,* excerpted from Amrit Sorli and Davide Fiscaletti. Special Theory of Relativity in a Three-dimensional Euclidean Space, *Physics Essays,* Vol. 25, No. 1 (March 2012), pp.141–143.

101 Carlo Rovelli. *The Order of Time,* Riverhead Books, New York (2018), pp.58–59.

102 Grigori Grabovoi. *Restoration of the Human Organism Through Concentration on Numbers,* RARE WARE Medienverlag Publishers (28 August 2011).

103 Michio Kaku. *Hyperspace: A Scientific Odyssey.* https://mkaku.org/home/articles/hyperspace-a-scientific-odyssey/

104 Michio Kaku. The Future of Humanity, *Talks at Google.* https://www.youtube.com/watch?v=eMxmDPDyQ7o

105 The Editors. *Collective Unconscious.* https://www.britannica.com/science/collective-unconscious

106 Hazel Guest. *The Origins of Transpersonal Psychology* (Sept. 1989). https://onlinelibrary.wiley.com/doi/pdf/10.1111/j.1752-0118.1989.tb01262.x

107 Ali Qadir and Titiana Tiaynen-Qadir. 'Towards an Imaginal Dialogue: Archetypal Symbols between Eastern Orthodox Christianity and Islam', *Approaching Religion,* Vol. 6 No. 2 (Dec. 2016), p.83.

108 John Keats. Letter to Benjamin Bailey (22 Nov. 1817). http://www.towernotes.co.uk/literature-notes-18_Keats_Odes_1.php

109 Janey Davies. 'The Hare Psychopathy Checklist with 20 Most Common Traits of a Psychopath', *Learning Mind* (accessed 11 August 2018). https://www.learning-mind.com/hare-psychopathy-checklist.

110 Lou Agosta. Empathy and Sympathy in Ethics, Internet *Encyclopaedia of Philosophy* (accessed 11 August 2018). https://www.iep.utm.edu/emp-symp/#SH3a.

111 Ibid.

112 Ibid.

113 Brihadaranyaka Upanishad 2.3.6.

114 Shankara. *Vivekachudamani.* Brahman. *The Best Upanishads Quotes,* Paramahansa Yogananda. http://yogananda.com.au/upa/Upanishads01.html

115 Plato. *Timaeus,* trans. Benjamin Jowett. http://classics.mit.edu/Plato/timaeus.html

116 Rupert Sheldrake, Terrence McKenna, and Ralph Abraham. *Chaos, Creativity and Consciousness,* Rochester, VT. Park Street Press (2001), pp.4, 20.

117 Alex Vikoulov. The Unified Field and the Quantum Nature of Consciousness,

#Ecstadelic (7 July 2016). https://www.ecstadelic.net/ecstadelic/the-unified-field-and-the-quantum-nature-of-consciousness.

118 Sourya Acharya and Samarth Shukla. Mirror Neurons: Enigma of the Metaphysical Modular Brain, *Journal of Natural Science, Biology and Medicine,* Vol. 3, No. 2 (July–December 2012), pp.118–24. https://www.ncbi.nlm.nih.gov/pmc/articles/PMC3510904.

119 Dr Phil Maffetone. *Sunlight: God for the Eyes as Well as the Brain.* https://philmaffetone.com/sun-and-brain/

120 Rick Strassman. *The Spirit Molecule,* Kindle, p.54.

121 Ibid. p.53.

122 Len Wisneski and Lucy Anderson (book review). The Scientific Basis of Integrative Medicine, *Evidence-based Complementary and Alternative Medicine,* Vol. 2, No. 2 (June 2005), pp.257–59. https://www.ncbi.nlm.nih.gov/pmc/articles/PMC1142191.

123 Stasia Bliss. NASA Confirms Superhuman Abilities Gained, *Liberty Voice* (29 May 2013). http://guardianlv.com/2013/05/nasa-confirms-super-human-abilities-gained.

124 Dr Edward Group DC, NP, DACBN, DCBCN, DABFM. *Everything You Wanted to Know About the Pineal Gland,* Global Healing Center (20 April 2016). https://www.globalhealingcenter.com/natural-health/everything-you-wanted-to-know-about-the-pineal-gland/

125 Ibid.

Through the Looking Glass – The Power of Identity

126 Frank Pajares. *But They Did Not Give Up,* University of Kentucky (accessed 11 August 2018). https://www.uky.edu/~eushe2/Pajares/OnFailingG.html.

127 Kendra Cherry, reviewed by Steven Gans MD. *What Are the Id, Ego and Superego? The Structural Model of the Personality.* https://www.verywellmind.com/the-id-ego-and-superego-2795951

128 C G Jung. *Collected Works of C G Jung,* Vol. 9, Part 2, trans. Gerhard Adler, R.F.C. Hull.

129 Khalid Jamil Rawat. *Instinct and Ego: Nietzsche's Perspective,* Vol. 2015(2), pp.61–70, E-LOGOS, University of Economics, Prague.

130 Sri Sathya Sai Baba. God, *Quotations.* http://www.saibaba.ws/quotes/god.htm

131 What Is the Ego? Lesson 331, Section 12, *A Course in Miracles,* Mill Valley, CA, Foundation for Inner Peace (2007).

132 Cathy Stinear, Applied Practical Neuroscientist. Lecture, *The Creative Mind,* Creative AI Conference, Wellington, NZ (July 2018).

133 Michael Pollen. *How to Change Your Mind: What the New Science of Psychedelics Teaches Us About Consciousness, Dying, Addiction, Depression, and Transcendence,* New York, Penguin Press (2018), p.305.

134 Ibid. p.304.

135 Felicity Callard, Jonathan Smallwood, and Daniel S. Marguiles. *Default Positions: How Neuroscience's Historical Legacy Has Hampered Investigation of the Resting Mind.* https://www.ncbi.nlm.nih.gov/pmc/articles/PMC3437462/

136 Spire Wellness Team. *Your Complete Guide to Holotropic Breathing Benefits and Techniques,* Spire Wellness blog (5 November 2017). https://blog.spire.io/2017/11/05/what-is-holotropic-breathing.

137 *What Is Rebirthing Breathwork?* Rebirthing Breathwork International

(accessed 11 August 2018). www.rebirthingbreathwork.com/2013/03/13/what-is-rebirthing-breathwork.

138 Everett L Shostrum, Producer and Director. *Abraham Maslow and Self Actualization* (1968 documentary). https://www.youtube.com/watch?v=7DOKZzbuJQA.

139 *Brain Structure May Be Root of Apathy*, University of Oxford Medical Sciences Division (13 Nov. 2015). https://www.medsci.ox.ac.uk/news/brain-structure-may-be-root-of-apathy. Also: Valerie Bonnell, et al. Individual Differences in Premotor Brain Systems Underlie Behavioral Apathy, *Cerebral Cortex*, Vol. 12, No. 2 (1 Feb. 2016). https://academic.oup.com/cercor/article/26/2/807/2367142.

140 Elizabeth W Dunn, et al. 'Spending Money on Others Promotes Happiness', *Science*, Vol. 319, No. 5870 (21 March 2008), pp.1687–8. http://science.sciencemag.org/content/319/5870/1687.

Humanity's Quantum Leap – The Power of the Presence

141 *I Am That*, Dialogues of Sri Nisargadatta Maharaj, trans. Maurice Frydman, Durham NC, Acorn Press (2012), p.1.

142 John White. *Resurrection and the Body of Light*. https://www.theosophical.org/publications/quest-magazine/42-publications/quest-magazine/1690-resurrection-and-the-body-of-light

143 His Eminence Dzogchen Khenpo Choga Rinpoche. *Rainbow Body Practitioner*. https://bodhiactivity.wordpress.com/2013/11/29/rainbow-body-practitioner/

144 John 10.34, *World English Bible* (Public Domain).

145 Exodus 3:14, *World English Bible* (Public Domain).

146 Ibid.

147 Meeting of Ramana Maharshi and Paramahansa Yogananda, *Arunachala Mystic blog* (16 Nov. 2014). http://arunachalamystic.blogspot.com/2014/11/meeting-of-ramana-maharsi-and.html.

148 *I Am That*, Dialogues of Sri Nisargadatta Maharaj, 10. Witnessing, p.27.

149 Ibid. 7. 'The Mind', p.18.

150 Ibid. 10. 'Witnessing', p.27.

151 Ibid. 100. 'Understanding Leads to Freedom', p.385.

152 Ibid. 86. 'The Unknown is the Home of the Real', p.329.

153 El Morya, The Bridge to Freedom Library, February 1962, quoted in *21 Essential Lessons, Vol. 1*, Compiled by Werner Schroeder, Ascended Master Teaching Foundation.

154 Ibid. p.1.

155 Physicist Fritjof Capra. *The Tao of Physics: An Exploration Between Modern Physics and Eastern Mysticism* (1975), Epilogue, p.305.

156 Maha Chohan. *Thomas Printz' Private Bulletin, Book 2*, Ascended Master Teaching Foundation, Kindle, locations 8105–8110.

157 Kuthumi. *Teachings for The New Golden Age*, pp.195–197, Ascended Master Teaching Foundation (2002).

158 Ibid.

159 Job 22:28, *World English Bible* (Public Domain).

160 *21 Essential Lessons, Vol. 1*. Compiled from the teachings of *The Bridge to Freedom* by Werner Schroeder, Ascended Masters Teaching Foundation (2008), p.91.

161 Ibid. p.95.

162 Matthew 17:20, *World English Bible* (Public Domain).

163 Scott Jeffrey. *How to Discover and Transmute Trapped Emotion.* https://scottjeffrey.com/repressed-emotions/

The Practical Application of Creative Alchemy

Phase One – Clearing Suppressed Emotion and Unleashing Power

165 *Review of the Germanic/German New Medicine of the Discoveries of Dr Ryke Geerd Ham.* http://www.newmedicine.ca/overview.phpHamer
166 Ibid.
167 Ibid.

Phase Two – Living Vision

168 'Song of Myself', *Leaves of Grass* (1892), Section 1. https://whitmanarchive.org/published/LG/1891/poems/27 or https://www.gutenberg.org/files/1322/1322-h/1322-h.htm
169 Ralph Waldo Emerson. 'Self-Reliance'. *Essays: First Series*, The Literature Page. www.literaturepage.com/read/emersonessays1-33.html
170 Ulrich Hoerni, Thomas Fischer and Bettina Kaufmann (Eds.) *The Art of C.G. Jung*, WW Norton & Company (2019), p.260.
171 Neel V Patel. *Retrocausality is the Key to Time Travel* (5 Jan. 2016). https://www.inverse.com/article/9896-retrocausality-is-the-key-to-time-travel-what-the-hell-is-retrocausality
172 Steve Nadis. 'Physicists Show Time Flows Asymmetrically at the Electron Level', *Discover.* http://discovermagazine.com/2013/june/02-physicists-show-time-flows-asymmetrically-at-the-electron-level
173 Carlo Rovelli. *The Order of Time*, Riverhead Books (2018), pp.194–195.
174 Special thanks to Dr Norma Milanovich for this technique and permission to share it here.
175 Henry Miller. *Stand Still Like the Hummingbird*, New Directions Paperbook.
176 IFL Science. *Scientists Discover Plants Have Brains That Tell Them When to Grow.* http://www.iflscience.com/plants-and-animals/scientists-discover-plants-brains-determine-grow/ And: *Scientists Discover Plant 'Brain' Controlling Plant Development*, University of Birmingham, Eureka Alert! Global Source for Science News. https://www.eurekalert.org/pub_releases/2017-06/uob-sdp053117.php
177 Suzanne Simard. TED Talk. *How Trees Talk to Each Other.* https://www.ted.com/talks/suzanne_simard_how_trees_talk_to_each_other
178 Ibid.

Phase Three – Connection to the Magic Presence

179 John 12:23, *World English Bible*, (Public Domain).
180 Swami Muktánanda. *Meditate, Happiness Lies Within You,* SYDA Foundation.
181 Beloved Maha Chohan. *Thomas Printz' Bulletin Book 2*, Ascended Masters Teaching Foundation (1995), p.504. Special thanks to Cathi Brandon.
182 Stephanie Sinclaire. *God's Theory of Creativity: an Odyssey*, available on Amazon.
183 Werner Schroder, Bridge of Freedom Library. Special thanks. www.ascendedmaster.org
184 Patricia Cota Robles. *Invoking the Violet Flame.* https://eraofpeace.org/pages/premium-content/?mc_cid=44fbfe45e2&mc_eid=a1e56a38a0#media-popup

Phase Four – Putting it All Together

185　Thomas Printz. *The Seven Mighty Elohim Speak On: The Seven Steps to Precipitation*, Ascended Masters Teaching Foundation (1986).

186　Werner Schroeder, The Law of Precipitation, p. 97, Ascended Master's Teaching Foundation.

187　Ibid.

188　Vishen Lakhiani. 'The Surprising Thing I Learned from Studying My Brain with Meditation Technology for 7 Days', *Mindvalley* (16 Jan. 2018). https://blog.mindvalley.com/studying-brain-with-meditation

Co-Creating a New World

190　Dan Shawbel and Walter Isaacson. 'What Can We Learn About Innovation from Leonardo da Vinci?' *Forbes*, (16 Oct. 2017). https://www.forbes.com/sites/danschawbel/2017/10/16/walter-isaacson-what-we-can-learn-about-innovation-from-leonardo-da-vinci/#14bc13ad3d6a

191　Jack Preston and Richard Branson. 'Take Risks, Don't Avoid Them', *Virgin* (1 June 2017). https://www.virgin.com/entrepreneur/richard-branson-take-risks-dont-avoid-them

192　Frederick R Barnard. *Printer's Ink* (Trade journal, 10 March 1927), p.114.

193　Elysa Fenenbock. Lecture, *Creative Leadership NZ*, Conference, Wellington, NZ (2017).

194　Ibid.

195　Gus Balbontin. Lecture, *Creative Leadership NZ*, Conference, Wellington, NZ (2017).

196　*About B Corps*, The B Corps Declaration of Interdependence. https://bcorporation.net/about-b-corps

197　Michelle Giddens. The Rise of B Corps Highlights the Emergence of a new Way of Doing Business, *Forbes*. https://www.forbes.com/sites/michelegiddens/2018/08/03/rise-of-b-corps-highlights-the-emergence-of-a-new-way-of-doing-business/

198　Fritjof Capra. *A Conceptual Framework for Ecological Economics Based on Systemic Principles of Life*, (11 Sept 2017). https://www.fritjofcapra.net/a-conceptual-framework-for-ecological-economics-based-on-systemic-principles-of-life/

199　Ibid.

200　Ram Singh. *Vipassana in Jails: An Historical Review*, Vipassana Research Institute. https://vridhamma.org/research/Vipassana-in-Jails-An-Historical-Review

201　'Gaia Theory: Model and Metaphor for the 21st Century', *Overview, Understanding Gaia Theory*. http://www.gaiatheory.org/overview/

202　Sir James Jeans. *The Mysterious Universe*, Cambridge University Press (1931), p.69.

203　Genesis 1:3, *KJV* (Public Domain).

204　Paul Gorman. A Man of Faith Who Has Seen the Light, *The Dominion Post* (4 May 2019), C2.

205　Interview in *The Observer* (25 Jan. 1931), p.17, column 3.

206　Self-Realization Fellowship. www.yogananda-srf.org

207　James Jeans. *The Mysterious Universe* (1930), Cambridge University Press Online (2010).

208　Richard Conn Henry. The Mental Universe, *Nature* 436:29, 2005.

BOOKS & RESOURCES OF INTEREST

Alexander, Stephon – *The Jazz of Physics: The Secret Link Between Music and the Structure of the Universe*

Barks, Coleman (trans.) – *The Essential Rumi*

Capra, Fritjof – *The Tao of Physics; The Turning Point; Uncommon Wisdom; The Ecology of Law, Toward a Legal System in Tune with Nature and Community*

Frydman, Maurice (trans.) – Sri Nisargadatta Maharaj, *I Am That*

Gershon, Michael MD – *The Second Brain*

Gibran, Kahlil – *The Prophet*

Gurdjieff, G.I. – *Meetings with Remarkable Men*

www.heartmath.com – The HeartMath Institute

Kahanov, Linda – *Rider Between the Worlds; The Tao of Equus: A Woman's Journey of Healing and Transformation Through the Way of the Horse*

King, Godfré Ray – *Unveiled Mysteries; The I AM Discourses* and other books from the Saint Germain Press.

Muktānanda, Swami – *Meditate: Happiness Lies Within You*

Murphy, Michael – *Golf in the Kingdom*

Pearce, Joseph Chilton – *The Biology of Transcendence; The Crack in the Cosmic Egg*

Pert, Candace – *Molecules of Emotion*

Pollan, Michael – *How to Change Your Mind*

Rovelli, Carlo – *The Order of Time*

Schelde, Karina – *Expression into Freedom*

Schroeder, Werner– *Electrons and the Elemental Kingdom*, and other books from the Bridge of Freedom library, www.ascendedmaster.com

van der Kolk, Bessel – *The Body Knows the Score*

Yogananda, Paramahansa – *Autobiography of a Yogi*

Yukteswar, Swami Sri – *The Holy Science*

PRAISE FOR
GOD'S THEORY OF CREATIVITY: AN ODYSSEY
BY STEPHANIE SINCLAIRE LIGHTSMITH
(FIRST PUBLISHED *AS THE SHORES OF GRACE – AN ODYSSEY*)

'It is a truly absorbing story, the beginning of an Odyssey, which from the very first lines breathes the spirit of Creativity. I am deeply impressed by your huge spectrum of receptivity. Every step is experienced viscerally, intellectually, emotionally and imaginatively, and pursued with relentless trust in the wonder that draws you on, however uncertain the moment.'

– Guyon Neutze, spiritual teacher and author of Dark Out of Darkness

'I have just finished reading *God's Theory of Creativity.* It is one of the most riveting books I have ever read. I am sharing its wisdom with friends and family all over the place. (I have already ordered another copy for a friend in New York.) It's a book whose time has surely come. My only question is, when is the next instalment coming out?'

– Cindy Laberge, lawyer

'*God's Theory of Creativity* charts a radical awakening. It records rare, mystical, transpersonal experiences as they are happening without the distance of memory, and dives deep into the shadow world of the unconscious, bringing up jewels from the mire. Part memoir – a raw and truthful retelling of an extreme childhood and its effects, part metaphysical odyssey and part unflinchingly honest healing journey, it offers rare glimpses of our glory and a moving, lyrical account of our humanity.'

– Dr Rebecca Weiss-Vlasic, clinical psychologist

'Reporting back from fantastic, dangerous and experimental territories with poetic elegance and pragmatism, Sinclaire permits us to enter a most private and sacred zone of human experience … a fearless companion on the journey to awareness.'

– Anne Hemenway, environmentalist, activist, writer

'Stephanie Sinclaire's *God's Theory of Creativity* is a vast and beautiful work. Richly illustrated with her own paintings depicting a prolonged mystical encounter, Sinclaire traverses realms I have only ever seen one other have the courage to do – and even then, only publish after his death. Permit me to explain. *God's Theory of Creativity* and Carl Jung's *Liber Novus* or *Red Book* are, for me, companion volumes – two extraordinary ventures into the realm of the Spirit and Soul, accompanied by avatars of the Divine. How was this extraordinary account of the dialogues and teachings on creativity and consciousness made available to both? I will explain with what I have come to know about Jung. At 38, he could no longer believe in or follow the psychoanalytic principles of his former mentor, Sigmund Freud. He set out on his own path, determined – like Sinclaire – to discover his own truth. He didn't believe human beings were simply a set of infantile complexes that reduce us to our basest instincts. Instead, Carl Jung believed in humanity's highest potential, and, even more so, once the shadow and the light within us are consciously integrated, that our true individuated self – what some would call our Higher or Divine Self – would both guide and shine through us. He once said, when asked if he believed in God, "I don't believe, I know." Sinclaire's title and her whole journey into this realm, proclaims this same truth.

'When Jung split with Freud, he let himself fall into what psychiatrists would now call a psychotic state. Each night he allowed visions to enter into his consciousness. He was accompanied by a wise guide called Philemon, encountering with angels, demons… and all the archetypes of the personal and collective unconscious. Each morning, as he integrated the light and the dark of his experience, he would enter normal life – the consulting room and his family duties. As he experienced these visions, so did he write and paint. His Red Book is the record of that journey beyond normal waking consciousness. He himself acknowledged that his life's work was really to unlock and understand what he recorded at that time.

'So, too, with Sinclaire's book. Not her theory of creativity, but God's. To read and be immersed in this journey to the heart of the numinous encounter, infinitely aware and creative, is to feel what few among us know – an intimate connection with the Divine. I believe that, like Jung, Sinclaire's life work has been to embody this "theory" as her lived offering, in loving service to others. I am forever grateful to have read such a sacred and intimate testimony. In its pages, you can touch God's robes, images and words taking you straight to the Source.'

– *Stefania Pietkiewicz, writer and life purpose counselor*

'This book is an intense, ethereal yet almost perilous journey into the creative experience. It is both nearly spiritually exclusive and still fascinatingly human. A fantastic journey for anyone searching their soul, contemplating their existence and most importantly, coming to terms and trying to understand their own artistry and creative being. It is lovingly accompanied by stunning artwork that in themselves inspire.'

– *Natasha Ragdale, journalist*

'Stephanie Sinclaire Lightsmith is the William Blake of our times.'

– *Mariela Durnhofer Rubolino, Argentinian translator* God's Theory of Creativity

ABOUT THE AUTHOR

Stephanie Sinclaire Lightsmith is an award-winning internationally exhibited artist, author, playwright, screenwriter, director and producer for theater and film, teacher of inspirational creativity and the founder of the healing and empowerment technique The Creative Alchemy Method™. In 2002 she was awarded the Golden Jubilee Award by Her Majesty Queen Elizabeth II for 'contribution to the arts and the pursuit of excellence' at the Royal Academy, London.

Stephanie has traveled the world learning from wisdom teachers, scientists and masters devoted to advancing human potential and also trained in many cutting-edge modalities addressing the integration of mind, body, emotion and spirit. In the mid-1990s she collated the fruits of her personal journey from trauma to healing and empowerment with her extensive training and her inspirational visionary experience recorded in the memoir *God's Theory of Creativity* to found The Creative Alchemy Method™. She has since facilitated it in four countries, in schools, at small and large retreats, one to one consultations, for businesses, with entrepreneurs and more. She is devoted to helping people reach their highest potential and reigniting humanity's creative spark at this crucial moment in our planetary evolution.

To stay in touch with Stephanie and be notified of courses and one to one consultations please visit www.creativealchemy.vision

If you enjoyed this book it would be very appreciated if you could share your thoughts and comments. Thank you so much. May your day be blessed and joyful.

GoodReads: https://www.goodreads.com/book/show/53642048-creative-alchemy

Amazon: https://www.amazon.com/Creative-Alchemy-Science-Miracles-Co-Create/dp/047350944X

CPSIA information can be obtained
at www.ICGtesting.com
Printed in the USA
BVHW060251190421
605283BV00016B/816

9 780473 564889